Artificial Intelligence *and* Machine Learning *for* Women's Health Issues

Artificial Intelligence *and* Machine Learning *for* Women's Health Issues

Edited by

Meenu Gupta
Department of Computer Science and Engineering, University Centre
of Research and Development, Chandigarh University, Punjab, India

D. Jude Hemanth
ECE Department, Karunya Institute of Technology and Sciences,
Coimbatore, India

ELSEVIER

ACADEMIC PRESS
An imprint of Elsevier

Academic Press is an imprint of Elsevier
125 London Wall, London EC2Y 5AS, United Kingdom
525 B Street, Suite 1650, San Diego, CA 92101, United States
50 Hampshire Street, 5th Floor, Cambridge, MA 02139, United States

ISBN 978-0-443-21889-7

For information on all Academic Press publications
visit our website at https://www.elsevier.com/books-and-journals

Publisher: Mara Conner
Acquisitions Editor: Sonnini Yura
Editorial Project Manager: Tom Mcarns
Production Project Manager: Kamesh R
Cover Designer: Christian Bilbow

Typeset by STRAIVE, India

Working together
to grow libraries in
developing countries

www.elsevier.com • www.bookaid.org

Contents

4. **Artificial intelligence approaches for ultrasound examination in pregnancy**

Somya Srivastava, Disha Mohini Pathak, and Gaurav Dubey

5. **Early assessment of pregnancy using machine learning**

Chander Prabha and Meenu Gupta

6. **Ensemble learning-based analysis of perinatal disorders in women**

Malvika Gupta, Puneet Garg, and Chetan Malik

7. Machine learning approaches to predict gestational diabetes in early pregnancy

Poonam Joshi, Sapna Rawat, Arpit Raj, and Vikash Jakhmola

8. Contribution of artificial intelligence to improving women's health in pregnancy

Gulafshan Parveen, Poonam Joshi, Yashika Uniyal, Haidar, and Sapna Rawat

17. Role of artificial intelligence and machine learning in women's health

Sapna Rawat, Poonam Joshi, Gulafshan Praveen, and Jyoti Saxena

Contributors

Numbers in parenthesis indicate the pages on which the authors' contributions begin.

Nagendiran Baskar (219), Department of Biochemistry, PSG College of Arts & Science, Coimbatore, Tamilnadu, India

Charvi (17), Computer Science and Engineering, SGT University, Gurugram, India

S. Divya (193), Centre for Advanced Data Science, Vellore Institute of Technology, Chennai, India

Kiran Dobhal (151,207), Uttaranchal Institute of Pharmaceutical Sciences, Uttaranchal University, Dehradun, Uttarakhand, India; College of Pharmacy, Shivalik Campus, Dehradun, Uttarakhand

Gaurav Dubey (57), KIET Group of Institution, Ghaziabad, India

Nidhi Gairola (207), Uttaranchal Institute of Pharmaceutical Sciences, Uttaranchal University, Dehradun, Uttarakhand, India

Puneet Garg (17,91,137), Computer Science and Engineering, St. Andrews Institute of Technology and Management, Farrukhnagar, Gurugram, Delhi NCR, India

Malvika Gupta (91), Department of CSE, ABES Engineering College, Ghaziabad, UP, India

Meenu Gupta (79), Department of Computer Science and Engineering, University Centre of Research and Development, Chandigarh University, Punjab, India

Haidar (121), Guru Nanak College of Pharmaceutical Sciences, Dehradun, Uttarakhand, India

H. Jude Immaculate (219), Department of Mathematics, Karunya Institute of Technology and Sciences, Coimbatore, Tamilnadu, India

Vikash Jakhmola (107,151,207), Uttaranchal Institute of Pharmaceutical Sciences, Uttaranchal University, Dehradun, Uttarakhand, India

Poonam Joshi (107,121,255), Uttaranchal Institute of Pharmaceutical Sciences, Uttaranchal University, Dehradun, Uttarakhand, India

Vijay Kumar (37), Manav Rachna International Institute of Research & Studies, Faridabad, Haryana, India

Chetan Malik (91,137), San Diego State University, San Diego, CA, United States

Medha Malik (137), Department of CSE, ABES Engineering College, Ghaziabad, UP, India

Durairaj Mohanapriya (219), Department of Computer Science, PSG College of Arts & Science, Coimbatore, Tamilnadu, India

Srishti Morris (163), Uttaranchal University, Dehradun, India

S. Naveen Venkatesh (193), School of Mechanical Engineering (SMEC), Vellore Institute of Technology, Chennai, India

Rahul Negi (173), Kamla Nehru College, University of Delhi, New Delhi, India

Kiran Pal (37), Delhi Institute of Tool Engineering, Delhi, India

Gulafshan Parveen (121), Guru Nanak College of Pharmaceutical Sciences, Dehradun, Uttarakhand, India

Disha Mohini Pathak (57), ABES Engineering College, Ghaziabad, India

Chander Prabha (1,79), Chitkara University Institute of Engineering and Technology, Chitkara University, Punjab, India

Gulafshan Praveen (255), Veer Madho Singh Bhandari Uttarakhand Technical University, Dehradun, Uttarakhand, India

Arpit Raj (107), Quantum University, Roorkee, India

Dipika Rawat (163), Uttaranchal University, Dehradun, India

Sapna Rawat (107,121,255), JBIT Group of Institution, Dehradun, Uttarakhand, India

Kunnathur Murugesan Sakthivel (219), Department of Biochemistry, PSG College of Arts & Science, Coimbatore, Tamilnadu, India

Akash Samanta (151), Uttaranchal Institute of Pharmaceutical Sciences, Uttaranchal University, Dehradun, Uttarakhand, India

Jyoti Saxena (255), JBIT Group of Institution, Dehradun, Uttarakhand, India

Mariappan Selvarathi (219), Department of Mathematics, Karunya Institute of Technology and Sciences, Coimbatore, Tamilnadu, India

Alka Singh (207), School of Pharmaceutical Sciences, Sardar Bhagwan Singh University, Dehradun, Uttarakhand

Vaishali Singh (235), XIM University, Bhubaneswar, India

Somya Srivastava (57), ABES Engineering College, Ghaziabad, India

V. Sugumaran (193), School of Mechanical Engineering (SMEC), Vellore Institute of Technology, Chennai, India

Jyotsana Suyal (151), Uttaranchal Institute of Pharmaceutical Sciences, Uttaranchal University, Dehradun, Uttarakhand, India

Yashika Uniyal (121), Guru Nanak College of Pharmaceutical Sciences, Dehradun, Uttarakhand, India

Shalu Verma (207), Uttaranchal Institute of Pharmaceutical Sciences, Uttaranchal University, Dehradun, Uttarakhand, India

Chapter 1

Role of artificial intelligence in gynecology and obstetrics

Chander Prabha

Chitkara University Institute of Engineering and Technology, Chitkara University, Punjab, India

1.1 Introduction

In the majority of medical disciplines over the last two decades, artificial intelligence (AI) usage has grown tremendously. There are two types of AI in medicine that can be applied: symbolic and nonsymbolic AI. Symbolic AI is based on ontologies and knowledge, whereas nonsymbolic AI is based on the usage of methods like machine learning (ML), artificial neural networks (ANNs), and so on. Generally, the usage of AI is spread across numerous domains of gynecology and obstetrics (GYN/OB) [1]. A few examples are gynecology surgery, assisted reproductive medicine, general obstetrics, and fetal ultrasound.

Numerous definitions exist for AI [2]. In general, AI is the use of complex algorithms that allow machines to perform cognitive functions, like decision-making and problem-solving. It can also be defined as a machine's capability to emulate intelligent human behavior. AI comprises four elements: computer vision, ML, natural language processing (NLP), and ANNs. ML makes use of complex datasets to draw inferences to make future predictions. ML outperforms logistic regression in predicting surgical infection via analysis of multiple data comprising diagnosis and treatment. ML is a part of AI that extracts useful information from a huge dataset. NLP is designed in such a way that it understands human language. Inspiration for ANNs comes from the biological nervous systems where neurons serve as computational units. In computer vision, images and videos are studied. For example, analyzing sleeve gastroscopy of laparoscopic videos in real time using computer vision achieved an accuracy of 92.8% [3].

Nowadays, the attempt to use refined AI clinically is more active. Newer technologies have been developed with regard to AI in medicine [4], for example, deep learning (DL) using ANN to analyze medical images. AI applications that use DL have surpassed magnetic resonance imaging (MRI), computed

Artificial Intelligence *and* Machine Learning *for* Women's Health Issues
https://doi.org/10.1016/B978-0-443-21889-7.00013-0

1

tomography (CT), pathology slides, and ultrasonography to, for example, diagnose or determine disease severity via endoscopy of intestinal images. AI-based video platforms are now being developed for commercialization.

AI provides support for taking clinical decisions in addition to assuring the quality of images. To detect GYN/OB-related diseases, ultrasonography is the most commonly used technique. Although AI is still in its infancy, it has shown its potential in examining and accurately predicting gestational age (GA), fetal structures, fetal positioning, and more. To make AI use in GYN/OB a reality, communication between sonographers and AI developers needs to be encouraged [5].

Furthermore, the rapid development of genetic engineering in in vitro fertilization (IVF) procedures increased the demand for AI to improve precision [6]. Conventional techniques, such as evidence synthesis and clinical trials, continue to be the most significant instruments in addressing health concerns among women. Nonetheless, the gray areas present in conventional approaches continue to be the cause of clinical practice's failure to provide suitable solutions [7]. As a result, AI is now an important part of life these days, as it is in medicine, particularly digital medicine.

This chapter describes and discusses AI technologies being developed and researched for application in GYN/OB. It also highlights recent advances in medicine using AI for early diagnosis of maternal-fetal conditions like abnormal fetal growth and preterm birth (PTB). It discusses the clinical benefits of AI and AI's role in diagnosis of GYN/OB conditions, IVF, obstetric ultrasound, pregnancy, PTB, and maternal-fetal conditions. It discusses the existing constraints and the future prospects of AI in GYN/OB, along with obstacles faced in medical procedures. Finally, the chapter presents concluding remarks.

1.2 Clinical benefits of GYN/OB via AI

The current application of AI in GYN/OB is addressed as the dominant tool in the identification of problems with pregnancy, PTB, and reviewing clinical interpretation discrepancies. These advantages help clinicians reduce rates of maternal and infant morbidity and mortality. AI serves as a powerful means for developing algorithms to detect women with asymptomatic shorter cervical lengths (CLs) and at risk of PTB [7]. Additionally, the considerable benefits of the enormous storage capacity of AI might include the identification of risks related to preterm labor via pervasive genomic data and multiomics, thus reducing complications during pregnancy, reducing operational time, and assisting surgeons in learning in a real-time environment.

1.3 AI in obstetrics

Uterine contractions and fetal heart conditions are monitored via cardiotocography (CTG), which is used for screening hypoxia development and is considered a major tool in intrapartum decision-making. However, the interpretation of

CTG is prone to human error due to the high variability in inter and intraobserver. Many studies on the prediction of fetal outcomes to aid in decision-making have been conducted by looking at CTG data. These studies demonstrated ML functionality in fetal monitoring to determine whether, during intrapartum care, cesarean section is mandatory or not [8].

AI applications are exceptional, as they address perennial problems in medical investigation and treatment. AI could supplement knowledge and assist medical professionals in making judgments in GYN/OB [9]. The understanding of fetal physiology and CTG could be aided by AI, limiting the negative effects of obstetrics. Due to higher intraobserver and interobserver fluctuation, CTG understanding may be susceptible to misinterpretation, thus leading to human errors. DL and ML may be used as supplements in fetal monitoring and logically endorses in determining the need for a cesarean section during intrapartum care. ML is an AI subfield that defines, in general, a machine's ability to imitate the intelligent behavior of humans.

In 2014, to identify CTG patterns (normal and abnormal) using latent class analysis, a huge dataset of CTG was classified into a random forest (RF) classifier of ML [10]. The obtained results of specificity and sensitivity for correctly interpreting CTG were 78% and 72%, respectively. In 2017, some parameters, including mother age, umbilical artery pH, CTG data, and base deficit and excess, were used for the classification of normal and cesarean section deliveries [11]. The parameters were processed using a subset of ML algorithms comprised of Fisher's linear discriminant (FLD) analysis, RF classification, and DL. In comparison, it was found that DL gives better results of specificity and sensitivity, with scores of 91% and 94%, respectively.

1.4   AI in IVF

The main challenge in IVF is the choice of a feasible embryo. This choice is necessary for predicting results, as it could lead to a healthy live birth and a short time to pregnancy. The author in [12] developed an ANN using four inputs (age, number of eggs recovered, freezing of embryo or not, and number of embryos transferred) with a predictive power of 59% in order to calculate the chances of successful IVF. Challenges here are many unknown factors that may result in unsuccessful IVF. To increase the chances of success, large datasets and computer vision techniques were used for constructing an ANN to improve predictive power. With traditional approaches, treatment for infertility is still a major concern. Therefore, AI-assisted IVF is an example of the increasing need for AI in GYN/OB for improved treatment success rates.

Few previous attempts at using AI techniques to predict normal fertilization, assess human oocytes, and analyze embryo development to blastocyst stages have previously been made. These methods have even examined implantation possibility in the pre and postpregnancy intervals using static oocyte images.

1.5 AI in obstetric ultrasound

Ultrasound is a common method utilized throughout the entire pregnancy process. Fetal growth can easily be determined along with the detection and treatment of disease [13]. Ultrasound provides high-quality images of fetal anatomy, providing detailed information for improved diagnostic accuracy. Presently, 2D and 3D imaging techniques are popularly used for measuring fetal structures, assessing organ functions, and diagnosing disease. However, during the early stage of pregnancy, due to involuntary fetal movements, the fetal structures are almost always clotted in the second and last trimester, resulting in misdiagnosis and difficulty in the examination. During the last two decades, to limit intra and interobservation variation in obtaining automatic measurements and improve diagnostic accuracy, AI-based assessments have been widely used. Three aspects are involved in improving the accuracy of obstetric ultrasound when AI is applied. These are identification of structure, standardized and automatic measurements, and diagnosis classification (Table 1.1).

The use of AI improves workflow and reduces the time of examination, as obstetric ultrasound is time-consuming. Related research on AI-aided techniques is still in the early stages. Much commercial software has been introduced and launched to provide precise measurements and high-resolution imaging for obstetric ultrasound. One major step in enacting the clinical applicability and validity of AI techniques is interdisciplinary communication. There has been discussion reviewing the applications, pros, and cons of AI in obstetric ultrasound [14]. Different phases are involved in pregnancy, thus AI-aided obstetric ultrasound is divided into the first-second-last trimester. The future perspective of AI in obstetric ultrasound includes telediagnosis service or telemedicine and virtual reality (VR) learning. The fusion of various disciplines in medicine will expedite the adaptation of AI techniques to clinical applications [15].

1.6 AI application in trimester (first)

With its capabilities, AI will make possible a far more efficient and effective system for health care. AI can help with prognosis and diagnosis, therapy optimization, and discovering new drugs. AI is rapidly gaining popularity in the discipline of GYN/OB. Most of the relevant studies made use of images from ultrasounds, but doctors also experimented with AI-based CTG evaluation for intrapartum fetal monitoring [16].

1.6.1 Automatic estimation of fetal growth and development

The first 3 months of pregnancy are critical for examining pregnancy complications like low birth rate, premature delivery, and so on. The traditional method of fetal growth examination is measuring crown-rump length using

TABLE 1.1 AI application in improving the accuracy of obstetric ultrasound.

Aspects	Trimester first	Trimester second and last
Identification of structure	Fetal limb detection	Facial structure recognition
		Abdominal organ recognition
		Ventricular volume and ventricular wall thickness calculation
Standardized and automatic measurements	Automatic measurement of crown-rump length and nuchal translucency (NT)	AC/HC estimation automatically
		Volume estimation of the fetal head and its internal structure
		Volume estimation of the fetal stomach and bladder
Diagnosis classification	Fetal development assessment	Maturity prediction of neurodevelopment
		Growth restriction diagnosis
		Craniocerebral malformation diagnosis
		Maturity assessment of fetal lung
		Any prediction of premature birth by cervical ultrasound
		Congenital heart disease diagnosis
		Fetal weight and gestational age (GA) estimation

2D ultrasonography [17]. However, this method sometimes fails to detect the difference between abnormal and normal fetuses in the first trimester. The volume measurement provides further information when done in 3D as compared to 2D [18]; however, it is time-consuming. Keeping this in mind, another semi-automatic method was introduced based on 3D image volume calculation [19]. It was based on point-of-interest detection and pixel extraction. As an experiment, this semiautomatic method was used in a 12-week-old fetus to calculate the fetal volume and identify the fetal contours to verify the persuasiveness for accuracy of 3D imaging. However, the method failed to identify erratic items, and some segmentation faults required manual intervention. In a further development, during the first trimester, a 3D convolutional neural network (CNN) was used to realize the multiple anatomical structures of the fetus and placenta as well as determine the GA.

In another study, a solution based on image processing was proposed for the detection of fetal limbs, biometry view estimation, biometry automatic measurement, and fetus segmentation. This method used 3D ultrasound. As the fetal appurtenances and fetus were closely related, the usage of these algorithms gives a comprehensive and systematic assessment [20].

1.6.2 Nuchal translucency (NT) thickness assessment

To detect chromosomal malformations in pregnancy, NT thickness measurement was done by 2D ultrasonography. NT is a measurement of the maximum thickness between the subcutaneous soft tissue and the fetal skin. This measurement requires expertise and a high degree of accuracy, and thus, measuring NT thickness generally requires numerous attempts. This is due to small fetal structure, frequent fetal movement, and low image quality. To reduce dimensional errors and decrease dimensional measuring difficulties, one study measured NT thickness via semiautomated tools. In the study, Ref. [21] measured the fetal NT thickness via a standard median sagittal image in combination with a DNN. Based on this combination, a model was constructed that automatically recognized by combining 3D image data with sagittal plane information. This model achieved an accuracy of 88.6%. Fig. 1.1 shows the NT thickness measurement by automatic segmentation of the nuchal membrane and the edge of soft tissue overlying the cervical spine using AI. The calculation of the minimum vertical distance between the two lines computes the largest as the NT measurement.

1.7 AI application in second and last trimesters

AI-powered diagnostic assistance tools have outperformed in various medical dimensions. However, due to a dearth of interpretive power in AI judgments

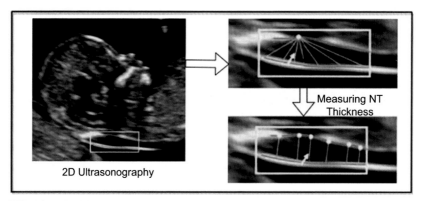

FIG. 1.1 NT thickness measurement via AI.

(known as the "black box" problem), the clinical implementation of AI is still difficult. This obstacle makes establishing rapport with medical practitioners difficult. Sakai et al. [22] used a novel DL-based understandable graph chart diagram illustration to support cardiac screening of the fetus via ultrasound, which has an inadequate rate of identification of congenital heart defects in its second-trimester stages due to the trouble in mastering the approach. Consequently, AI screening efficacy during the second and last trimesters for identifying pregnant women via diagram portrayal increases from 96% to 97.50% for professionals and from 82% to 89% for peers [22].

To detect fetal abnormalities, treat prenatal conditions, and assess fetal growth and development via prenatal ultrasonography is critical for reducing mortality rates. The most complicated organ to evaluate is the fetal brain. When AI-aided ultrasound is applied to detect fetal brain abnormalities, various aspects need to be considered. These aspects include biometry of the fetal head, cranial capacity measurement, fetal head circumference (HC) segmentation, and features (normal or abnormal) pertaining to ultrasound [23,24].

The plane of fetal ultrasound varies quickly, as fetal movement necessitates an algorithm that is capable of evaluating image motion in real time [25]. A prenatal diagnostic ultrasound AI conduct system (PAICS) has been suggested for the creation of an AI system to image multiple abnormal fetal intracranial structures in standardized sonographic reference planes to screen for any congenital central nervous system (CNS) malformations.

1.8 AI application in gynecology

The use of AI in gynecology has grown dramatically in the modern era. As diseases become more prevalent, there is a greater need to improve the process of disease detection. Several investigators have used the beneficial effects of AI in predicting diseases in this context [26,27].

Table 1.2 shows the applications of AI-based models in gynecology for the identification of IVF, uterine sarcoma, endometrial carcinoma, and cervical intraepithelial neoplasia, and in the evolution of anticancer drugs.

1.9 Application of ML in the early examination of maternal-fetal conditions

PTB can cause neonatal death [35]. The prediction of PTB during the first and second trimesters might help to improve the outcome of pregnancy. As a result, research has focused on developing an efficient forecasting model for PTB based on ANNs. Several studies supported the use of sonographic examinations to predict PTB during the first trimester of pregnancy.

Pelvic floor dysfunction (PFD) is yet another gynecological illness. Fecal incontinence, urinary loss, pelvic organ prolapse, and sexual dysfunction are

TABLE 1.2 Application of AI-based models in gynecology.

Disease	Author and year	Need	Methodology	Conclusion
Endometrial carcinoma	Dong H.C. et al. (2020) [28]	The important prognostic factors that could be used to stratify risk are DMI, LVSI, and FIGO. As a result, a thorough, noninvasive screening approach is required that is capable of predicting tumor stage in clinical practice	This study used a dataset of 137 women with endometrial cancer with an aggregate diameter of >1 cm who had 1.5 MRIs performed prior to hysterectomy. The appearance of the texture was then evaluated using commercially available software, and the adjacent region of focus was delimited manually	The testing results are similar to those of the most skilled radiologists
	Ueno Y. (2017) [29]	The development of computer-aided design (CAD) strategies in images from ultrasounds is required to improve the diagnostic precision of ultrasound examination	Three different AI-based strategies (ANN, CART, LR) were used. The dataset included postmenopausal women with endometrial widths of 5 mm having sporadic bleeding from the vagina. The diagnostic accuracy of three strategies was compared	AI, particularly DL, with improved specificity and sensitivity is a useful and effective computational tool that might be used in the healthcare industry to substantially enhance the health of the general population
	Malek M. (2019) [30]	MRI based on texture feature used in MRI and CAD might be paired with AI to distinguish clinical pathologic prognosticators prior to treatment	Endometrial cancer myometrial invasion in MRIs was detected using CNN and DL	Results suggest that AI can assist medical professionals (radiologists) in detecting the stage-1, myometrial invasion depth
Uterine sarcoma	Xue Y. (2019) [31]	The views on a semi-CAD structure stationed on PWI would be helpful to differentiate leiomyomas and uterine sarcoma	Radiologists extracted seven parameters from each ROI to characterize the contrast agent dynamic. As a result, the input was fed into the DT ensemble classifier, which categorized lesions as benign malignant uterine leiomyoma of uterine sarcoma	The suggested path is effective for obtaining an appealing discriminative power that could be used in conjunction with traditional MRI to distinguish sarcomas from myomas

Anticancer drug	Arsalan M. (2022) [32]	AI is critical in drug development research	Two gallic acid derivatives were synthesized as amides and esters. Spectral data such as MMR, FTIR, and MS were used to characterize new compounds. The A2780 cell line was used to test these compounds in vitro. For revealing flawless anticancer activities, an innovative method with the inclusion of the Hill function was used. SVM was also used in conjunction with pharmacodynamics modeling to validate the efficacy of anticancer drugs	The results revealed that all compounds tested were biocompatible and had substantial anticancer properties contrary to ovarian cancer cells (A2780). It has been suggested that the approach could effectively predict the in vitro result of the originating compounds produced in preliminary
IVF	Xie H.N. (2020) [13]	IVF prognosis is a critical accomplishment in aiding imitation, significantly benefiting sterile couples, neighborhoods, and health systems	Integrated omics and AI methods were developed to improve treatment success rates. First, the demographic parameters, subfertile living of couples, as well as IVF cycle features, were recorded. Then, the biomarkers, transcriptomics, and metabolomics were evaluated and measured, and deep ML analysis of oocyte, embryo, and sperm was performed. The developed ANN system improved the success rate of IVF cycles by increasing accuracy and objectivity as opposed to a traditional strategy	It was anticipated that the development of an easy and cost-effective system would be used in IVF units, especially after the massive assessment of transcriptomics and metabolomics, which lowers cost of procedures and would be available for daily use
	Siristatidis C. (2021) [33]	An embryologist evaluated blastocyst viability by manually microscopically examining its parts, including inner cell mass, zona pellucida, the blastocoel, and trophectoderm	SSS-Net was used to detect blastocyst components for embryological analysis	The proposed method could be used to validate the morphological properties of blastocysts for an effective IVF procedure
CIN	Sheril L. (2019) [34]	The detection and automatic segmentation of an erratic region in a cervical image is critical in determining the presence of cervical cancer	To distinguish abnormal cervical tissue, an ANN system was implemented based on an available spectral database	The researchers discovered that the rate of hemoglobin in unit tissue raised during the proliferation of cells thus raising the light absorption coefficient. It was proved that the proposed system can distinguish CIN from normal cervical tissue

the most prevalent signs and symptoms. Studies have investigated the potential advantages of the use of ultrasound and training for rehabilitation based on an AI algorithm in the recovery of postpartum pelvic organ prolapse.

As a result, AI algorithms have a positive impact on the interpretation of ultrasonic visuals. Lower abdomen rehabilitation training improved postpartum nursing in patients with genital prolapse. Furthermore, numerous wearable gadgets, such as smart rings and smartwatches, are proficient at recording rudimentary physiological measures like body temperature, heartrate fluctuations and average heartrate, oxygen saturation, and blood pressure levels. They also monitor additional behavioral indicators such as sleep quality, duration of sleep, and patient location and motion. The process of observing physiological parameters has obvious advantages for precisely determining initial pregnancy-related conditions such as preeclampsia and gestational hypertension [36].

Heartrate surveillance, activity monitoring, and diagnosis practices are the most common applications of AI and digital technology during pregnancy. However, incorporating specialized tools could lead to greater results from the moment of conception to the postpartum period. For example, the better understanding the metamorphosis of pregnancy (BUMP) study is a longitudinal study of the feasibility of using digital resources and ANN implementation to monitor and understand prepregnancy and postpregnancy symptoms.

The use of ANN and RF results in similar or better performance measures than conventional methods like logistic regression [37]. In an earlier study conducted on participants at Duke Medical Center, the independent variables (attributes) were obstetric variables consisting of race, age, religion, education, marital status, and so on. The dependent variable (class) was PTB. The area under curve (AUC) of the ANN was 0.68, which was better than that of logistic regression (0.66). The authors in Ref. [38] compared ML methods for the early prediction of PTB. They included a total of 596 obstetric patient records and used variable importance to find major predictors of PTB. In terms of accuracy, the results of ANN (0.911) were analogous to logistic regression (0.918) and RF (0.891). The variable importance in ANN focused on diabetes mellitus, hypertension, and prior cone biopsy. However, in RF, more emphasis was on prior PTB, age, and CL. The results indicate that the direct factors must be agreed with the indirect factors. AI can recognize the most viable embryos and oocytes with an excellent likelihood of pregnancy by combining with ANNs to derive appearance descriptors from embryo or oocyte pictures.

1.10 Existing constraints and future prospects

There are several limitations to current research on AI for the initial detection of maternal-fetal conditions. First, most research used a cross-sectional approach, and thus, data enhancement with a longitudinal design would improve AI's performance in the maternal-fetal domain. Second, these studies did not examine the potential mediating accouterments of predictors for the initial detection of maternal-fetal circumstances. Third, the literature is inadequate and thus extended research on feasible connections among maternal-fetal circumstances and gastroesophageal reflux disease is required [39]. Fourth, not many studies have been conducted on the ethical issues surrounding the use of AI in maternal-fetal conditions. However, if the present scenario continues, the circumstances are expected to evolve, and skillful working in maternal-fetal health care will need to fully address these problems [40]. Finally, no translational or basic research on maternal-fetal conditions has been carried out using AI.

Using big data in future studies might be an ideal approach. In addition, combining various DL approaches for numerous kinds of data on maternal-fetal conditions could result in deeper and more significant clinical repercussions. Also, whether categories that are binary (no, yes) for maternal-fetal conditions can evolve into more precise categories will be a fascinating area for research in the future.

The present research looked at the latest developments in the use of AI for the early detection of multiple maternal-fetal conditions like abnormal fetal growth and PTB. According to the findings of this research, AI has the potential to be an effective and harmless decision-support tool for the initial detection of maternal-fetal conditions [41].

1.11   Research gap

The differential assessment of malignant and benign adnexal masses via ultrasound examinations continues to be the most difficult challenge that clinical employees often encounter in gynecologic practice. The transvaginal ultrasound was thought to be the best method to detect differences and was considered the initial imaging technique. Furthermore, results of transvaginal ultrasound have primarily depended on the experience of the investigator, implying that the primary fault of this tool is that its accurate interpretation is primarily based on the one-sided impact of investigating clinical sonographers [42]. Furthermore, in patients with coincide benign polypoid tumors or leiomyomas, AI was more likely to result in incorrect interpretations [43,44].

1.12   Discussion on obstacles of AI

As there are numerous probable advantages to incorporating AI into medical procedures, there are also numerous obstacles and uncertainties that may cause concern. The fundamental constraints on the durability of these techniques are the quality and quantity of data utilized for creating the models.

Significant progress has been made on the usage of AI in GYN/OB; however, the success rate and ubiquity of various methods still require extended investigation [45–47]. Furthermore, as algorithms are constantly modified and optimized, the ideas and reasoning underlying these approaches should be understood not only by those who develop algorithms but also by healthcare professionals. The AI-based ultrasound techniques for obstetrics have gradually gained prominence, thus playing an important part in educational and social services. The online fetal ultrasound facility in medical care could connect the specialists in the fetal clinical center with a distant obstetric unit, allowing for high-trait ultrasound examination and expert advice while also reducing travel time. The technology has already been demonstrated to be useful in overseas appointments and consultations. Obstetric ultrasound is still inaccessible in rural regions and developing countries, and thus, diagnostic and telemedicine technologies could increase the acceptance of obstetric ultrasound examinations in situations with limited resources. As a result, more research in this area is required. AI is still in its infancy and is not yet a only strategy. Medical professionals remain encouraged to use AI appropriately to optimize and hypothesize its application in clinical practice [48,49].

1.13 Conclusion

AI appears to be an interesting tool in GYN/OB for addressing several long-standing issues. This chapter suggests that AI may enhance knowledge and help clinicians to make decisions in areas of GYN/OB. AI has the potential to improve fetal physiology and CTG interpretation, potentially reducing conflicting events in obstetrics. In the field of gynecology, AI can lay out the intricacy of gynecological cancer molecular biology and thus assist the perception of personalized medicine.

In medical and academic disciplines, the use of AI technologies has recently increased exponentially in maternal-fetal medicine, and it has become more prevalent in screening and making helpful medical decisions. Even though conventional statistical techniques have made convincing developments in determining disease causes, pathophysiology, evaluation, prognosis prediction, and treatment, AI technologies are emerging as subsequent methodologies. As maternal-fetal medicine entails two lives (the mother and fetus), more stringent ethical practices and standards will be required for the use of AI technologies; thus, greater focus should be paid to the quality, amount, and accuracy of data as well as the research inputs in this direction.

References

[1] L. Sarno, et al., Use of artificial intelligence in obstetrics: not quite ready for prime time, Am. J. Obstetr. Gynecol. MFM (2022) 100792.

[2] P. Iftikhar, et al., Artificial intelligence: a new paradigm in obstetrics and gynecology research and clinical practice, Cureus 12 (2) (2020) e7124, https://doi.org/10.7759/cureus.7124.

[3] A. Danial Hashimoto, et al., Computer vision analysis of intraoperative video: automated recognition of operative steps in laparoscopic sleeve gastrectomy, Ann. Surg. 270 (3) (2019) 414–421, https://doi.org/10.1097/SLA.0000000000003460.

 [4] M.M. Seval, B. Varlı, Current developments in artificial intelligence from obstetrics and gynaecology to urogynaecology, Front. Med. Sec. Obstetr. Gynecol. 10 (2023), https://doi.org/10.3389/fmed.2023.1098205.

 [5] G.S. Desai, Artificial intelligence: the future of obstetrics and gynecology, J. Obstet. Gynecol. India 68 (2018) 326–327, https://doi.org/10.1007/s13224-018-1118-4.

 [6] Y.-T. Shen, et al., Artificial intelligence in ultrasound, Eur. J. Radiol. 139 (2021) 109717.

 [7] M.R. Hassan, S. Al-Insaif, M.I. Hossain, J. Kamruzzaman, A machine learning approach for prediction of pregnancy outcome following IVF treatment, Neural Comput. Applic. 32 (7) (2020) 2283–2297.

 [8] S. Gennady, et al., The state of and prospects for the introduction of artificial intelligence technologies in obstetric and gynecological practice, Obstetr. Gynecol. 2 (2021) 5–12, https://doi.org/10.18565/aig.202L2.5-12.

 [9] E.I. Emin, et al., Artificial intelligence in obstetrics and gynaecology: is this the way forward? In Vivo (Athens, Greece) 33 (5) (2019) 1547–1551, https://doi.org/10.21873/invivo.11635.

[10] C. Sundar, M. Chitradevi, G. Geetharamani, Incapable of identifying suspicious records in CTG data using ANN based machine learning techniques, J. Sci. Ind. Res. 73 (8) (2014) 510–516.

[11] Y. Zhang, Z. Zhao, Fetal state assessment based on cardiotocography parameters using PCA and AdaBoost, in: Proc. 10th Int. Congr. Image Signal Process., Biomed. Eng. Informat, CISP-BMEI, 2017, pp. 1–6.

[12] P. Vogiatzi, A. Pouliakis, C. Siristatidis, An artificial neural network for the prediction of assisted reproduction outcome, J. Assist. Reprod. Genet. 36 (7) (2019) 1441–1448, https://doi.org/10.1007/s10815-019-01498-7.

[13] H.N. Xie, N. Wang, et al., Using deep-learning algorithms to classify fetal brain ultrasound images as normal or abnormal, Ultrasound Obstet. Gynecol. 56 (2020) 579–587, https://doi.org/10.1002/uog.21967.

[14] Z. Chen, et al., Artificial intelligence in obstetric ultrasound: an update and future applications, Front. Med. (2021) 1431.

[15] K.H. Ahn, K.-S. Lee, Artificial intelligence in obstetrics, Obstet. Gynecol. Sci. 65 (2) (2022) 113–124, https://doi.org/10.5468/ogs.21234. eISSN 2287-8580.

[16] Z. Hoodbhoy, M. Noman, A. Shaque, A. Nasim, D. Chowdhury, B. Hasan, Use of machine learning algorithms for prediction of fetal risk using cardiotocographic data, Int. J. Appl. Basic Med. Res. 9 (4) (2019) 226–230.

[17] D. Shigemi, S. Yamaguchi, S. Aso, H. Yasunaga, Predictive model for macrosomia using maternal parameters without sonography information, J. Maternal-Fetal Neonatal Med. 32 (22) (2019) 3859–3863.

[18] N.A. Smeets, N.A. Dvinskikh, B. Winkens, S.G. Oei, A new semi-automated method for fetal volume measurements with three-dimensional ultrasound: preliminary results, Prenat. Diagn. 32 (2012) 770–776, https://doi.org/10.1002/pd.3900.

[19] X. Yang, L. Yu, et al., Towards automated semantic segmentation in prenatal volumetric ultrasound, IEEE Trans. Med. Imaging 38 (2019) 180–193, https://doi.org/10.1109/TMI.2018.2858779.

[20] H. Ryou, M. Yaqub, A. Cavallaro, et al., Automated 3D ultrasound image analysis for first trimester assessment of fetal health, Phys. Med. Biol. 64 (2019) 185010, https://doi.org/10.1088/1361-6560/ab3ad1.

[21] J. Moratalla, K. Pintoffl, R. Minekawa, et al., Semi-automated system for measurement of nuchal translucency thickness, Ultrasound Obstet. Gynecol. 36 (2010) B412–B416, https://doi.org/10.1002/uog.7737.

[22] A. Sakai, M. Komatsu, et al., Medical professional enhancement using explainable artificial intelligence in fetal cardiac ultrasound screening, Biomedicine 10 (2022) 551, https://doi.org/10.3390/biomedicines10030551.

[23] X. Yang, X. Wang, Y. Wang, et al., Hybrid attention for automatic segmentation of whole fetal head in prenatal ultrasound volumes, Comput. Methods Prog. Biomed. 194 (2020) 105519, https://doi.org/10.1016/j.cmpb.2020.105519.

[24] H.K. Kang, A. Kaur, A. Dhiman, Menopause-specific quality of life of rural women, Indian J. Commun. Med. 46 (2) (2021) 273–276, https://doi.org/10.4103/ijcm.IJCM_665_20.

[25] I.D. Pluym, Y. Afshar, K. Holliman, L. Kwan, Accuracy of three-dimensional automated ultrasound imaging of biometric measurements of the fetal brain, Ultrasound Obstet. Gynecol. 57 (2020) 798–803, https://doi.org/10.1002/uog.22171.

[26] G. Delanerolle, et al., Artificial intelligence: a rapid case for advancement in the personalization of gynaecology/obstetric and mental health care, Women Health 17 (2021) 17455065211018111.

[27] H.K. Kang, A. Kaur, S. Saini, R. Ahmad, W.S. Kaur, Pregnancy-related health information-seeking behavior of rural women of selected villages of North India, Asian Women 38 (2) (2022) 45–64, https://doi.org/10.14431/aw.2022.6.38.2.45.

[28] H.C. Dong, H.K. Dong, M.H. Yu, et al., Using deep learning with convolutional neural network approach to identify the invasion depth of endometrial cancer in myometrium using MR images: a pilot study, Int. J. Environ. Res. Public Health 17 (2020) 5993, https://doi.org/10.3390/ijerph17165993.

[29] Y. Ueno, B. Forghani, R. Forghani, et al., Endometrial carcinoma: MR imaging-based texture model for preoperative risk stratification—a preliminary analysis, Radiology 284 (2017) 748–757, https://doi.org/10.1148/radiol.2017161950.

[30] M. Malek, M. Gity, A. Alidoosti, et al., A machine learning approach for distinguishing uterine sarcoma from leiomyomas based on perfusion-weighted MRI parameters, Eur. J. Radiol. 110 (2019) 203–211, https://doi.org/10.1016/j.ejrad.2018.11.009.

[31] Y. Xue, Y. Zhao, L. Yao, W. Li, Z. Qian, Development of diffuse reflectance spectroscopy detection and analysis system for cervical cancer, Chinese 43 (3) (2019) 157–161, https://doi.org/10.3969/j.issn.1671-7104.2019.03.001. 31184068.

[32] M. Arsalan, A. Haider, J. Choi, K.R. Park, Detecting blastocyst components by artificial intelligence for human embryological analysis to improve success rate of in vitro fertilization, J Pers Med 12 (2022) 124, https://doi.org/10.3390/jpm12020124.

[33] C. Siristatidis, S. Stavros, A. Drakeley, et al., Omics and artificial intelligence to improve in vitro fertilization (IVF) success: a proposed protocol, Diagnostics (Basel) 11 (2021) 743, https://doi.org/10.3390/diagnostics11050743.

[34] L. Sherin, A. Sohail, S. Shujaat, Time-dependent AI-modeling of the anticancer efficacy of synthesized gallic acid analogues, Comput. Biol. Chem. 79 (2019) 137–146, https://doi.org/10.1016/j.compbiolchem.2019.02.004.

[35] N. Baños, A. Perez-Moreno, et al., Quantitative analysis of cervical texture by ultrasound in mid-pregnancy and association with spontaneous preterm birth, Ultrasound Obstet. Gynecol. 51 (2018) 637–643, https://doi.org/10.1002/uog.17525.

[36] N. Prema, M. Pushpalatha, Machine learning approach for preterm birth prediction based on maternal chronic conditions, in: Emerging Research in Electronics, Computer Science and Technology. Proceedings of International Conference, ICERECT 2018, Springer, New York, 2019, pp. 581–588.

[37] K.S. Lee, K.H. Ahn, Artificial neural network analysis of spontaneous preterm labor and birth and its major determinants, J. Korean Med. Sci. 34 (16) (2019) e128, https://doi.org/10.3346/jkms.2019.34.e128. 31020816. PMC6484180.

[38] K.S. Lee, I.S. Song, E.S. Kim, K.H. Ahn, Determinants of spontaneous preterm labor and birth including gastroesophageal reflux disease and periodontitis, J. Korean Med. Sci. 35 (14) (2020) e105, https://doi.org/10.3346/jkms.2020.35.e105. 32281316. PMC7152528.

[39] H.B. Kann, et al., Artificial intelligence in oncology: current applications and future directions, Oncology (Williston Park, NY) 33 (2) (2019) 46–53.

[40] A.S. Shazly, et al., Introduction to machine learning in obstetrics and gynecology, Obstet. Gynecol. 139 (4) (2022) 669–679.

[41] G. Rizzo, M.E. Pietrolucci, A. Capponi, et al., Exploring the role of artificial intelligence in the study of fetal heart, Int. J. Card. Imaging 38 (2022) 1017–1019, https://doi.org/10.1007/s10554-022-02588-x.

[42] H.Y. Kim, G.J. Cho, H.S. Kwon, Applications of artificial intelligence in obstetrics, Ultrasonography 42 (1) (2022) 2–9.

[43] L.S.E. Eriksson, E. Epstein, et al., Ultrasound-based risk model for preoperative prediction of lymph-node metastases in women with endometrial cancer: model-development study, Ultrasound Obstet. Gynecol. 56 (2020) 443–452, https://doi.org/10.1002/uog.21950.

[44] R. Kaur, R. Kumar, M. Gupta, Predicting risk of obesity and meal planning to reduce the obese in adulthood using artificial intelligence, Endocrine 78 (3) (2022) 458–469.

[45] J. Balayla, G. Shrem, Use of artificial intelligence (AI) in the interpretation of intrapartum fetal heart rate (FHR) tracings: a systematic review and meta-analysis, Arch. Gynecol. Obstet. 300 (2019) 7–14, https://doi.org/10.1007/s00404-019-05151-7.

[46] R. Kaur, R. Kumar, M. Gupta, Food image-based nutritional management system to overcome polycystic ovary syndrome using DeepLearning: a systematic review, Int. J. Image Graph. 2350043 (2022).

[47] R. Kaur, R. Kumar, M. Gupta, Food image-based diet recommendation framework to overcome PCOS problem in women using deep convolutional neural network, Comput. Electr. Eng. 103 (2022) 108298.

[48] N.S. Malani, et al., A comprehensive review of the role of artificial intelligence in obstetrics and gynecology, Cureus 15 (2023) 2.

[49] R. Kaur, R. Kumar, M. Gupta, Deep neural network for food image classification and nutrient identification: a systematic review, Rev. Endocr. Metab. Disord. (2023) 1–21.

Chapter 2

Prediction of female pregnancy complication using artificial intelligence

Charvi[a] and Puneet Garg[b]

[a]Computer Science and Engineering, SGT University, Gurugram, India, [b]Computer Science and Engineering, St. Andrews Institute of Technology and Management, Farrukhnagar, Gurugram, Delhi NCR, India

2.1 Introduction

In recent decades, there has been a significant surge of interest in artificial intelligence (AI) across multiple domains, including health services and medical management. In the medical services, the term "prediction" typically refers to diagnosing the probability of an undetected health condition. AI research in medicine has been particularly focused on predicting various health-related issues such as diabetes, cardiovascular diseases (CVDs), obesity, cancer, pregnancy-related disorders, and more.

In pregnancy health care, various AI algorithms are used for the prevention and projection of multiple diseases, such as gestational diabetes mellitus (GDM), and machine learning (ML) is increasingly being used as well. AI methods are predicated on rule-based decision-making. The biopsychosocial diagnostic model [1] emphasizes the importance of considering not only biological factors but also psychological and social factors in predicting and diagnosing health issues, as shown in Fig. 2.1. This means that prediction models should work with diverse data from various sources, including not only the medical field but also inputs from patients, healthcare professionals, smartphones, and remote sensing devices. Additionally, a person's psychological data is considered a relevant data source that can be used to develop AI techniques for detecting emotional states. However, the variety of available data also raises questions about how the collection and consumption of the data is done, such as what data to collect, how rapidly to collect it, and how to amalgamate all various types of data to produce and revise health predictions. Taking a holistic

Artificial Intelligence *and* Machine Learning *for* Women's Health Issues
https://doi.org/10.1016/B978-0-443-21889-7.00001-4

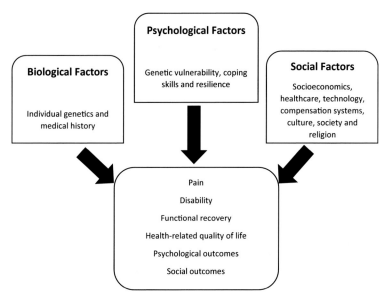

FIG. 2.1 Biopsychosocial diagnostic model.

approach to health care can provide more accurate and comprehensive predictions and diagnoses. Using rule-based decision-making, we can mirror a clinic's environment effectively. The most used algorithm in this context is MYCIN, which predicts appropriate therapies for various bacterial infections. It is a type of expert system used in women's health with a preterm, rule-based birth risk predictor. Utilizing state-of-the-art technology, this system determines the risk of preterm birth using various diagnostic codes during the pregnancy being evaluated. Restructuring is the only way for these kinds of programs to function better. AI's ML application learning possible without being explicitly programmed. An artificial neural network (ANN), a well-liked ML technique, is intended to mirror how organic neural networks process data.

The use of AI in women's health remains limited despite the development of the first AI tool for preterm birth 25 years ago and the first expert health-based AI system 45 years ago. Two recent studies highlighted the need for new state-of-the-art algorithms to check pharmacological effects in pregnant and lactating women, as most medicines are given to and taken by pregnant women without any clinical trials. Women who are pregnant or recently gave birth have not participated in research because of their vulnerability, practical and ethical concerns, and the intricacy of the subject matter. There is still a need for more research in this area, including the use of AI. An organization that conducts research on pregnant lactating women has outlined several methods for identifying and acknowledging gaps in knowledge and research regarding drug use in this population. AI and ML could be used to assess various data methodologies discovered in clinical data to provide more accurate and trustworthy results accountable to outcomes relating to pregnant women and fetuses, the design of innovative animal models for trials that ensure accurate results, and the generation of alarms for patients and doctors [2].

Understanding the fetal and maternal consequences of taking pharmacological medications during pregnancy poses a significant challenge for various research methods due to the limited availability of data based on controlled trials, which are randomized in humans. Typically, studies rely on populations that show high risk, such as those with HIV, those who have given preterm birth, and so on. However, there is a research gap regarding medication use during pregnancy for the general pregnant population. To address this issue, researchers must rely on retrospective outcome studies using data collected during routine clinical care. Although animal models provide data, there are significant loopholes in converting this data into the context of humans. Hence, existing information should be rearranged to tackle clinical questions, for which the use of AI techniques can be particularly powerful.

AI has the potential to transform health care, and predicting pregnancy difficulties is one area where it holds a lot of promise. AI can examine vast quantities of patient data and find patterns that might foretell possible issues including GDM, preeclampsia, and premature labor with the use of ML algorithms. AI may offer early warnings and allow healthcare practitioners to take preventative actions to lower the risk of pregnancy difficulties by assessing numerous parameters such as mother age, preexisting conditions, and prior pregnancy history. This makes AI a valuable tool in maternal and child health since it may result in better health outcomes for both the mother and the infant.

With an emphasis on pregnant women, this chapter reviews current research on AI approaches as used during pregnancy. It also discusses areas in which AI has room to grow.

2.2 Review

Within this review, we present various types of AI and explore their applications, current research, advantages, drawbacks, and future prospects.

2.2.1 Monitoring of fetal heart and surveillance of pregnancy

Fetal heart rate (FHR) monitoring is a crucial tool used by healthcare physicians to observe the fetus and identify any high-risk issues. It offers crucial details on baseline FHR, variability, acceleration, deceleration, the strength of uterine contractions, and any alterations in the FHR pattern. Nowadays, by examining cardiotocography (CTG) results and making predictions about likely outcomes, AI is being utilized to monitor FHR throughout childbirth. According to a recent study [3], implementing AI in obstetrics, particularly in antepartum surveillance, offers significant advantages. It might result in more reliable and consistent output for each study, less variation in how obstetricians interpret intrapartum surveillance, and eventually decrease neonatal and mother morbidity. AI systems can also offer supporting data in situations where predicted negative results may result in legal action. When a computer-aided fetal evaluator (CAFE) was used to examine CTG data, for instance, the findings demonstrated that the AI system was able to interpret the data at a level equivalent to that of subject-matter experts and was also able to detect errors. However, there were significant limitations to the study, such as disagreements in the expert interpretation of some data, which made it challenging to confirm the correctness of the system. Consequently, more study is needed to increase the data interpretation flexibility as well as the precision and efficacy of AI systems used for FHR monitoring.

2.2.2 Gestational diabetes mellitus (GDM)

GDM is an expensive and uncomfortable condition during pregnancy. AI, however, has the potential to screen for GDM and thus improve outcomes. Beyond 24 weeks of gestation, the current screening recommendations call for a glucose challenge test and a second diagnostic examination. The United States Preventative Services Task Force [4] concluded that there was not enough data to evaluate women before 24 weeks of pregnancy. To avoid issues like neonatal hypoglycemia, big for gestational age babies, and an increase in caesarean deliveries, among others, it is crucial to screen for GDM. While caesarean deliveries are more frequent in GDM, the cost of the outcomes is also higher; however, this cost may be less if a useful screening test is available.

To create a more useful and economical screening tool, Polak et al. [5] conducted research that screened for GDM using an AI calculator. The recommended online calculator is built on a multilayer perceptron ANN, which may be used by both patients and medical professionals. Long-term healthcare costs may be decreased by AI technology, which is also useful for daily tasks.

2.2.3 Labor in preterm birth

The prenatal outputs in women with short cervix length were investigated by Singh et al., [6] who employed AI, atrial fibrillation (AF) proteomics, and metabolomics together with imaging, demographic, and clinical data. The kind of AI used in this research, called deep learning (DL), is well suited for biological system analyses that include multiomics since it can handle larger data volumes and has more neural networks than other types of AI.

Short cervical length is now the biggest risk factor for preterm birth, yet a lot of women with this problem bring their pregnancies to term. Amniocentesis is used in several institutions to examine inflammatory and infectious processes to examine additional factors that might endanger women with short cervix lengths. In this investigation, the AF of the participants has been further examined for metabolomics to find possible novel biomarkers connected to premature delivery, in addition to helping doctors better stratify women at pre-term birth risk beyond the existing indicators of risk.

With this tool, medical professionals may more efficiently direct the treatment they provide for their patients, offering suggestions as necessary. The study found that DL was a better method for prediction and that additional research is necessary to explore the relationship between AF omics and premature shortening of the cervix, despite the fact that the study's small sample size made it difficult to draw firm conclusions. In these situations, the information will be used to guide patient treatment.

2.2.4 Parturition

Complications might arise if labor begins too early or too late, leading to a pre-term or postterm pregnancy. On study used AI to build gene circuit diagrams, which assisted in sifting through voluminous data to find pertinent details on myometrial activation. Nevertheless, since the study was carried out on guinea pigs, further investigation is required to confirm the findings in humans. Norwitz et al. [7] recommended more research to fully understand the genes involved in human parturition and confirm the results before applying AI to solve parturition-related concerns. In addition, factors like pregnancy problems must be considered in order to comprehend how they could affect myometrial gene expression. This research may help treat conditions like preterm labor and postterm pregnancy by giving information on the genes that become active during labor, hence decreasing the perinatal morbidity and death associated with these issues.

2.2.5 In vitro fertilization (IVF)

In research by Guh et al. [8], mining of data and AI was utilized to develop a computerized technique that could predict pregnancy outcomes while using IVF. Data mining is the process of obtaining valuable information from large databases, identifying patterns within them, and identifying new crucial elements that may have an impact on outcomes. The researchers developed an intelligent model that used GA-based and DT learning approaches to help physicians forecast results and alter IVF therapy based on particular patient features. Information was taken from IVF patient data using this model. Because the model only used data from one IVF clinic, it has limits; however, integrating data from several facilities may increase accuracy and represent a bigger population. ANN algorithms and learning vector quantizers, according to previous study, may be used to predict IVF outcomes more precisely. Another application of AI in IVF is to identify the most viable oocytes and embryos. An AI system was developed that combined texture descriptors extracted from pictures utilizing a local binary pattern with an ANN to objectively and noninvasively choose the best oocytes or embryos. Even in countries where selecting healthy embryos based on sex is illegal, the technology underscores the advantages of doing so.

2.2.6 Screening of cancer

Patients with ovarian cancer are given prognoses using neural network models. Many histologies, heterogeneous neoplasms, and patient presentations, such as the stage at which the tumor is growing, make ovarian cancer a complicated illness. According to Enshaei et al. [9], their constructed ANN system has a 97% accuracy rate when predicting survival. This technique may be able to provide patients with a reliable prognosis. Moreover, Norwitz et al. [7] created AI software that is more accurate than existing techniques at predicting prognosis and selecting the best course of therapy for ovarian cancer patients. Poor long-term survival statistics for advanced ovarian cancer point to the need for more specialized treatments.

2.2.7 Surgery in the field of gynecology

Another area in which AI has been applied is gynecological surgery via the use of surgical robots. Robot-assisted surgery has become increasingly common in gynecology, particularly for complex procedures such as hysterectomies. AI algorithms can help improve the precision of robotic surgery by analyzing patient data to provide better control and navigation of the surgical tools. For example, the da Vinci surgical system uses AI to filter out hand tremors and stabilize the instruments being used. This allows for more precise and accurate movements during surgery and can potentially reduce complications and improve outcomes.

Overall, the use of AI in gynecology has the potential to improve patient outcomes and enhance the efficiency and accuracy of various diagnostic and treatment methods. While there are still limitations and challenges to the implementation of AI in clinical practice, continued research and development in this field will likely lead to further advancements in gynecological care.

2.2.8 Discussion

It is important to note that while AI and ML have the potential to improve healthcare outcomes, they should not replace the role of healthcare providers. Clinicians should still be responsible for making final decisions based on AI-generated data and ensuring that patients receive the best possible care. Additionally, there is a need for continued training and education of healthcare professionals on the appropriate use and interpretation of AI-generated data.

Another potential downside of AI in health care is the issue of data privacy and security. As more patient data is collected and analyzed by AI algorithms, there is a risk of data breaches and unauthorized access to sensitive patient information. Therefore, appropriate measures need to be put in place to protect patient privacy and ensure data security.

Here are some important points to consider regarding the prediction of pregnancy complications using AI:

- Pregnancy problems must be identified early in order to improve both the health of both the mother and the fetus.
- AI systems can examine a lot of patient data to find trends and anticipate possible issues.
- Early warnings may be generated by analyzing variables including mother age, preexisting conditions, and prior pregnancies.
- Healthcare professionals should take precautions by anticipating issues, including gestational diabetes, preeclampsia, and premature labor.
- AI can change health care and enhance mother and child health outcomes.
- AI may assist healthcare professionals in making better judgments and giving pregnant patients individualized care.
- Research is still required to guarantee the usefulness and safety of AI's application in health care, which is still in its infancy.
- The use of AI in health care must also take ethical and legal issues into account.

Overall, the use of AI in the prediction of pregnancy difficulties has the potential to enhance the health outcomes of both mothers and their children. However, to guarantee that it can be used in a way that is both safe and effective, further study and careful consideration of ethical and legal considerations are required.

In conclusion, while AI has significant potential to enhance the outcomes of health care, there are a number of challenges relating to data privacy and security that need to be carefully explored and handled. These include the necessity for continuous clinical engagement.

2.3 Machine learning applications of maternal-fetal issues

Within this section, we describe different applications of ML in the field of maternal-fetal health.

2.3.1 Preterm birth

ANNs and random forests (RFs) have reportedly performed as well as or better than conventional methods like logistic regression (LR) when it comes to the early identification of preterm delivery. A recent study [10–13] used data from the Duke University Medical Center, which comprised 19,970 individuals and covered the period from January 1, 1988 to June 1, 1997. ANNs surpassed decision trees and LR in terms of performance, each with an area under curve (AUC) of 0.67.

Another recent study by Lee et al. [14] compared common ML techniques for the early identification of preterm delivery in a rare comparison. The study cohort included 596 obstetric patients who were admitted to Seoul Korea's Anam Hospital from March 27, 2014 to August 21, 2018. Using variable relevance, significant predictors of preterm birth were found. The ANN accuracy (0.9111) was comparable to that of RF (0.9178) and LR (0.9111). (0.8915). The ANN's variable importance results showed that diabetes mellitus, prior cone biopsy, and hypertension were important predictors of preterm birth, but the RF gave greater weight to cervical length, age, and past preterm delivery. The results of the former investigation were consistent with those of earlier studies, which recommended shifting the emphasis from direct to indirect causes of premature birth.

2.3.2 Abnormal growth in fetal

In research by Fung et al. [15], projected fetal weight was predicted using ultrasound measurements and ML methods. The Magee Obstetric Maternal & Infant (MOMI) Database from the University of Pittsburgh Medical Center (UPMC), USA, provided data on around 25,000 obstetric patients and their infants. The research included five independent variables, including infant gender, parity, 1-min/5-min Apgar scores, and gestational age at delivery. Fetal weight estimation served as the dependent variable. With a root mean squared error of 96.4, 104.7, and 127.3, respectively, the RF technique outperformed the decision tree and linear regression methods. This study is one of the first to predict estimated fetal weight using clinical and sonographic indicators and ML.

In research by Lee et al. [16], ML was used to estimate the body mass index of newborns utilizing 64 ultrasound maternal-delivery factors. The research comprised 3159 mother-newborn couples from a multicenter retrospective study by the Korean Society of Ultrasonography in Obstetrics and Gynecology. The body mass index of the infant was significantly predicted using the ultrasound maternal-delivery data. ANNs with one, two, and three hidden layers were outperformed by linear regression (2.0735) and RF (2.1604) in terms of mean squared error. According to the research, week 36 and beyond is the optimal timespan for ultrasound measures, and the abdominal circumference, estimated fetal weight, gestational age, and the mother's body mass index after delivery are the greatest predictors of the body mass index of newborns. This research was the first to calculate the body mass index of infants using ML.

2.4 Pregnancy and artificial intelligence

Even though some researchers have created emotional data-based prediction models for pregnancy health and wellbeing, a significant body of literature contends that a pregnant woman's emotional state might influence unfavorable pregnancy circumstances and outcomes, such as preeclampsia. Preeclampsia has been linked to several factors, including emotional stress and high blood pressure, demanding living circumstances, sleep issues, irregular circadian rhythms, obesity, and hereditary history of the condition. Preeclampsia, c-sections, low birth weight, and premature birth are just a few of the pregnancy issues that have been associated with anxiety and melancholy [17]. Determining a pregnant woman's likelihood of experiencing difficulties during pregnancy requires taking anxiety and sadness into account. As said, pinpointing pregnant women at risk for prenatal depression and anxiety can enable supportive and preventative measures. It is crucial to treat emotional wellbeing as part of comprehensive prenatal care given the severe negative effects of pregnancy on a woman's psychological health.

Pregnancy is a complex event that can have a significant impact on a woman's psychological health. Physiological changes that occur during pregnancy can be difficult to adjust to, both physically and mentally. Additionally, the pursuit of maternal and fetal wellbeing can require significant lifestyle changes and medical supervision, which can also affect a woman's mental state. Furthermore, the fear or worry of developing a pregnancy-related illness can further exacerbate mental health issues. While psychological problems can be a result of other underlying issues, they can also increase the likelihood of pregnancy complications.

A higher risk of developing hypertension conditions like preeclampsia has been linked to mental problems during pregnancy, sleep issues, and stressful life events, according to various research. Preeclampsia and nonvaginal births are only two examples of the negative pregnancy outcomes that anxiety and depression throughout pregnancy have been linked to. Anxiety and depression in the first trimester of pregnancy were shown to be linked to an elevated risk for preeclampsia in a prospective study of 623 pregnant women. Another study comprising 1371 pregnant women found a connection between maternal mental health and hypertension problems. Anxiety during pregnancy may be a risk factor for preeclampsia, with a greater incidence of anxiety seen in the preeclampsia group, according to case–control research including 300 pregnant women.

Research by examining the impact of fear about the start of bad health during early pregnancy on the pregnant woman's health, while not being directly connected to depression or anxiety. They discovered that worrying throughout the first trimester might directly affect how soon dangerous pregnancy consequences like preeclampsia will manifest. The study's authors hypothesized that any therapies and activities targeted at assisting the expectant mother manage her worries about pregnancy-related health issues could be advantageous for both the mother and child.

It is important to remember that stress during pregnancy may potentially have long-term effects on the behavior and development of the child. According to National Institutes of Health, there is a greater chance that children whose mothers had higher levels of stress during pregnancy could later face emotional and behavioral issues, including anxiety and attention deficit hyperactivity disorder (ADHD). According to different research, being exposed to stress during pregnancy was linked to changes in a child's brain development, notably in regions connected to attention and cognitive control.

Overall, these findings suggest that emotional distress during pregnancy can have a significant impact on both the mother and baby's health and wellbeing, both in the short term and in the long term. It is important for healthcare providers to screen for and address emotional distress during pregnancy, and for pregnant women to seek support and resources to manage their stress and anxiety.

Pregnant women have faced several difficulties because of the COVID-19 epidemic, which elevated stress levels and caused issues with mental health. Pregnant women are less likely to attend prenatal care visits in person or schedule them online because of measures used to contain the virus, such as social isolation and lockdowns, which have hampered support from family and friends. Lack of funds and an increased likelihood of interpersonal relationships and domestic violence are two additional risk factors for mental health problems during the epidemic. Maternal stress has been associated with a number of potentially detrimental outcomes for both mothers and infants, including an elevated risk of depressive symptoms, anxiety, suicidal thoughts, and inadequate mother-infant bonding.

It is essential to proactively create suitable ways to reduce stress throughout the epidemic. Several approaches have been suggested, including Web-based psychological support and therapeutic therapies, virtual online consultations and counseling, and routine depression and anxiety screening in obstetric settings. Furthermore, it has been proposed that providing psychological first aid and adequate risk communication are helpful in reducing adverse outcomes for pregnant mothers and their fetuses. Generally, it is crucial to keep an eye out for new dangers to unborn children and pregnant women during emergencies.

Additionally, the use of AI in perinatal mental health can provide valuable support to healthcare professionals in identifying and managing mental health disorders during pregnancy and the postpartum period. AI-based tools can help in early detection of mental health problems, monitoring the patient's emotional state, and providing personalized recommendations for managing stress and anxiety. For example, AI-based chatbots and virtual assistants can provide round-the-clock support to pregnant women, answering their questions and providing emotional support when needed.

Moreover, AI can assist in the integration of the biopsychosocial model into perinatal care, by analyzing a vast amount of data from different sources, including medical records, social media, wearable devices, and mobile applications. This can provide a more comprehensive understanding of the patient's health status and enable healthcare professionals to develop personalized care plans that take into account both biological and psychosocial factors.

Here are some crucial points to consider regarding the use of AI in pregnancy:

- AI has the potential to completely change prenatal care by offering individualized and more precise evaluations of pregnancy risks and problems.
- AI systems can examine vast volumes of data and spot patterns that are difficult for people to see, which enables early issue identification.
- Risk ratings that can be used to identify women who are more likely to have issues can be created using predictive analytics.
- Real-time access to patient data made possible by AI may also assist healthcare professionals in making more educated judgments, which can result in improved treatment strategies.
- Wearables with AI capabilities can also monitor the health of the fetus and spot possible issues before they become serious, potentially improving results for both mother and child.
- The use of AI during pregnancy does, however, bring up ethical issues related to privacy, prejudice, and algorithm openness.
- In addition to making sure healthcare personnel are adequately educated to utilize new technologies, it is crucial to guarantee that AI is used ethically and openly.
- To properly comprehend the possible advantages and hazards of employing AI during pregnancy and to decide how to effectively incorporate this technology into standard prenatal care, further study is required.

2.5 Understand maternal and fetal health outcomes from pharmacologic through recent uses of artificial intelligence

AI methods can be particularly powerful in repurposing existing data to answer important clinical questions related to medication use during pregnancy. Researchers must rely on retrospective outcome studies utilizing data collected during normal clinical treatment since there is a lack of research on drug usage during pregnancy in the general community. AI techniques can assist in processing this data to produce valid and trustworthy conclusions about maternal and fetal outcomes. Also, the best animal models for trials that confirm retroactive results discovered in healthcare records or other sources might be created using AI or ML techniques. Finally, AI techniques might be employed to promptly inform doctors and their patients about particular information pertaining to pregnant or nursing women.

2.5.1 Pregnancy and animal models

Proteomics was employed in a study by Cox et al. [18] to separate the preeclampsia patients into several subgroups based on genetic variations. To account for contamination of subcellular compartments, the researchers applied

Bayes Net ML to convert data from enhanced trophoblasts in mice to human preeclampsia placenta data. The study discovered that the three subgroups of preeclampsia in humans may be distinguished using maternal molecular abnormalities from placentas. This knowledge of the variety in preeclampsia ethology may be useful for customizing diagnostic and therapeutic strategies for preeclampsia patients.

2.5.2 Technologies based on assisted reproduction

Additionally, AI methods have also been applied to optimize assisted reproductive technology (ART) treatment by determining the best timing for embryo transfer. Wang et al. [19] developed a DL algorithm to predict the timing of embryo transfer that would result in the highest pregnancy success rate. The algorithm was trained on data from more than 6000 IVF cycles and was able to accurately predict the optimal timing for embryo transfer for individual patients.

Furthermore, AI methods have also been applied to understand the risk of miscarriage in ART. Jurisica et al. [20] developed an ML model that predicts the risk of miscarriage based on maternal age, number of previous miscarriages, and other clinical factors. The model was able to accurately predict the risk of miscarriage in a large cohort of women undergoing ART treatment.

Overall, AI methods have shown great promise in improving the success of ART treatment by predicting pregnancy outcomes, optimizing treatment, and understanding the risk of miscarriage.

Mora-Sa'nchez et al. [21] proposed a methodology that utilizes human leukocyte antigen haplotypes to predict the likelihood of recurrent miscarriage in couples compared to those with successful pregnancies. Forecasting pregnancy success is still challenging, and the crucial characteristics are typically missing from current models. Pharmacological treatments may be taken into account by certain models, although they may not be regarded as a crucial component. As a result, future studies in ART and ML could incorporate further variables, such as the kind of ovulation induction therapy employed.

2.5.3 Toxicology in development

Jelovsek et al. [22] developed a method to gather recommendations for AI application from a panel of experts in order to detect developmental toxicologic risks. The cumulative rule set was analyzed to classify each rule as either a confidence rule or an important rule after expert interviews and the gathering of their rules. Nonetheless, there was considerable disagreement among specialists over how to classify the rules, suggesting that the boundaries of the categorization were not clear. They identified six factors that influence expert judgment of whether there is a developmental toxicologic hazard: results of human and animal studies, presence of the active compound in humans, similarity of the compound's physical structure and mechanism of action to known toxicants, and whether there is a mutagen or direct cytotoxic agent. To expand rule set coverage, pharmacologic concepts must be taken into consideration.

2.5.4 Pregnancy and management of chronic disease

Preexisting diabetes, chronic hypertension, chronic respiratory conditions, and drug use disorders are the most common chronic illnesses among pregnant women in the United States. Research on the influence of chronic illness on pregnancy outcomes is scarce, however. According to one study, systemic lupus erythematosus (SLE) is linked to poor obstetric outcomes, such as fetal loss, in pregnant women. A clinical decision support system (CDSS) was created by Paydar et al. [23] employing two ANNs to forecast the pregnancy outcomes for pregnant SLE patients, namely spontaneous abortion or live delivery. The LR model's important features, which comprised characteristics from pharmacological interventions and lab test features, were used to train the ANNs. The MLP network was shown to be the most accurate after tenfold cross-validation, with a prediction accuracy of 91%.

Chronic illness can impact fertility and pregnancy outcomes as well as maternal and fetal health. To mitigate potential negative outcomes, it is recommended that preconception care includes screening, evidence-based care, and condition-specific management of chronic diseases.

2.5.5 Disease based on induced pregnancy

In addition to GDM, gestational hypertension disorders such as preeclampsia and eclampsia can also have severe consequences for maternal and fetal health. Preeclampsia is a dangerous pregnancy complication that may cause premature delivery and low birth weight. It is characterized by high blood pressure and organ damage, often to the liver and kidneys. Eclampsia is a severe preeclampsia complication that may cause seizures or convulsions that are potentially fatal for both the mother and the baby. For the best results for both the mother and the fetus, early diagnosis and therapy of these diseases are crucial.

ANNs were created by Polak et al. [5] to examine the correlation between demographic parameters and the risk of GDM. In comparison to LR, the accuracy of the ANN model was higher, correctly predicting 70.02% of true positive diagnoses as compared to LR's accuracy of 56.05%. In a separate study, researchers looked for potential GDM cases in pregnant women using an ANN-based technique called radial basis function network (RBFNetwork). An F-measure of 0.77 and a precision of 79.03% were obtained in this research. To aid hospital management, the RBFNetwork, which is based on ANNs in business intelligence, was developed. The authors contend that collaboration between expectant mothers, medical professionals, and healthcare administration personnel is necessary to reduce the occurrence of GDM.

CDSS can be used to manage the treatment of patients with GDM due to the need for increased monitoring. Applications of CDSS can enhance patient participation, nutrition advice, therapy planning, and home monitoring. To investigate home monitoring and therapy planning for GDM, a CDSS was created that combines qualitative and quantitative reasoning. The DIABNET system was developed to be used in patient visits, offering clinicians help by suggesting qualitative dietary changes and quantitative adjustments to insulin medication. In 92% of the situations where therapeutic adjustment was required, according to a system evaluation, the recommended alterations were approved as legitimate by expert assessors.

2.5.6   Labor anesthesia

Neuraxial labor analgesia, such as epidural anesthesia, is commonly used in the United States during childbirth to reduce pain, with a usage rate of 36%–81%. According to an algorithm created by Yu et al. [24] ultrasound images acquired from the lumbar spine of pregnant women automatically detect the bone/interspinous region in the transverse plane. The interspinous and bone pictures were identified with a high success rate of 92% on the test set using a support vector machine (SVM) model, and the ultrasound picture attributes were retrieved by the approach employing template matching and midline detection. In 45 of the 46 situations when the approach was evaluated using ultrasound video feeds, the spot was successfully identified. There is no research on anesthesia dose, even though this study concentrates on determining the best location for epidural anesthesia.

2.5.7   Delivery mode

Muzaffarabad, Jammu, and Kashmir conducted research [4] to identify the risk variables for female caesarean sections. They employed 23 components and 488 patients to categorize the participants, using different factors. Five classifiers—RF, linear discriminant analysis (LDA), SVM, Naive Bayes (NB), and k-nearest neighbor (*k*-NN)—were subjected to tenfold cross-validation in the study. The greatest performance, with the highest precision, accuracy, and recall, was determined to be RF. The results of the research showed that the predicted birth mode is highly influenced by the mother's age and the mode of the most recent birth. Although the study took into account medication as a characteristic, the researchers did not find it to be a crucial component in their model.

Various studies have focused on using ML techniques to enhance diabetes prediction in women diagnosed with GDM postpartum. GDM is linked to an elevated risk of metabolic syndrome and CVD, as well as a sevenfold increased risk of developing type 2 diabetes (T2D). Lin et al. [25] looked at the Artificial Immune Recognition System's (AIRS) capacity to predict the progression of DM after GDM. This supervised learning technique is motivated by the processes of both the natural and artificial immune systems. In AIRS, training and test data are seen as antigens, causing B-cells to produce artificial recognition balls (ARBs). With the same number of resources, these ARBs compete with one another, and those with more resources have a greater likelihood of creating mutant offspring that will be advantageous to the system. Wang et al. [19] developed a novel feature selection method that integrates the EM algorithm with the nearest neighbor classifier in order to successfully predict T2D in Taiwanese women after a GDM pregnancy. A convolutional neural network (CNN), an ML technique, was utilized by Meenakshi and Maragatham to assess the likelihood that T2D may eventually develop in women with GDM.

2.5.8 Pregnancy and pharmacology

To enhance patient care and provide drug information, expert systems have been employed along with patient counseling support. A prototype knowledge system was proposed by Swart et al. [26] in 1994, which prioritized information items based on their level of importance. Researchers have suggested new techniques to make rule-based system creation and maintenance easier. For instance, an employed ML was used to forecast the fetal toxicity of medications used during the whole course of pregnancy. To learn from medications that were known to be either hazardous or safe for the baby, researchers used the RF method. By analyzing the chemical makeup of the medications and determining whether they targeted a known Mendelian disease gene or a vitamin gene, the model was able to predict the fetal toxicity of pharmaceuticals that fell into the intermediate or unknown toxicity categories. A manual review procedure that validated the significance of Mendelian illness genes and the therapeutic potential of targeting them was used to create the subsequent ML technique.

AI may be used to anticipate how the usage of pharmaceuticals during pregnancy will affect mother and fetal health outcomes. AI systems may find trends and anticipate future adverse effects related to pharmacologic medicines by evaluating vast volumes of patient data. AI may also assist medical professionals in making better choices about the timing and dose of medications during pregnancy in order to enhance patient outcomes. As a result of recent developments in AI, prediction models that can identify fetal discomfort and forecast preterm labor have been created, enabling early treatments. In general, AI has the potential to enhance maternal and fetal health outcomes by providing accurate and individualized evaluations of pharmaceutical usage during pregnancy and by anticipating probable adverse events.

2.6 Maternal and fetal health outcomes based future application of artificial intelligence

Giving pregnant women and their medical professionals crucial information about how medications affect both the mother and fetus should be a priority as AI is developed in medicine, especially in the field of women's health. In doing so, it would enable both patients and doctors to make better informed decisions. One such AI technique that can aid in improved scientific insight extraction from current information is ML. Although there have been initiatives to improve pharmacovigilance and drug development, few have concentrated on enhancing medication safety during pregnancy.

2.6.1 Opportunities based on translational human research

We will be better equipped to improve the design of animal trials to confirm our assumptions in a more accurate and trustworthy manner if we have correct information on the possible damage a medicine may bring to the mother or the fetus. ML techniques can also be leveraged to aid and optimize the design of animal experiments tailored to the specific clinical questions and outcomes expected. For instance, since pigs have skin that is more similar to that of humans, ML can be utilized to assist in the design of animal studies.

2.6.2 Drug dosage optimization

During pregnancy, physiological changes can affect how drugs are metabolized in the body. This is especially important for the management of chronic diseases during pregnancy. There is a dearth of research on utilizing ML to guide medicine administration during pregnancy, despite some studies on treating chronic conditions like SLE during pregnancy. Nonetheless, initiatives have been undertaken in other fields including radiation oncology and warfarin dosage. There is still a vacuum in our understanding of the clinical implications of pregnancy-related alterations in medication metabolism for both the mother and fetus.

2.6.3 Clinical decision support opportunities

AI techniques can be utilized to advise expectant mothers and their doctors when the science has been created and confirmed. To provide the physician with the information they need to make the best possible clinical decisions, clinical workflow optimization must be optimized using AI technologies. For the clinical workflow process to be optimized, information timeliness is essential. Today's clinical decision support alerting systems often depend on antiquated AI "expert" system methods that utilize hardcoded decision trees. This process may be improved by using AI-powered, dynamic rule-based decision tree approaches that constantly improve the information offered to physicians by learning from users. Decision-making algorithms may be strengthened with more sophisticated AI approaches so that physicians and their patients get the relevant information at the appropriate moment.

2.7   Limitations

Most studies reviewed in this chapter employed cross-validation, but there was limited use of external validation. This lack of external validation makes it difficult to generalize the findings to a broader population. A further drawback is the small number of studies that have examined the use of AI to pharmaceutical therapies. Several of the papers that were eliminated focused on improving prenatal surveillance and screening for chromosomal and congenital abnormalities rather than stressing the crucial work in these areas. This emphasizes the significance of using AI and ML techniques in various areas of maternal reproductive health care. As a previous review on this subject has already been published, our study narrowed its attention to pregnancy and avoided looking at the function of AI in studies of lactation. Given its significance, it would be crucial to examine the potential of AI in pharmaceutical effects on breastfeeding in a subsequent review.

2.8   Conclusion

Our review showcases how AI has been employed to tackle the issue of pharmacological exposure during pregnancy, covering all stages from preconception to postnatal care. In particular, for pharmacological exposures relevant to pregnancy, our data shows a considerable gap in the application of AI approaches to enhance translational studies between animals and humans. All categories, however, have received insufficient attention on the relationship between pharmaceutical exposure and pregnancy. Consequently, using cutting-edge AI techniques is crucial for future research to better understand the effects of pharmacological drug exposure on the mother and fetus. By applying AI techniques to other scenarios of pregnancy, maternal, and fetal health, such as nursing, we may provide important information for the crucial study that will further explore the impacts of pharmacology on pregnancy. In order to increase the system's flexibility and eliminate bias in algorithm creation, further research is needed. This will enable the system to incorporate new medical knowledge when new technologies are created. Healthcare workers should take the required measures of safety to check that the analyzation is trustworthy and relevant. It is critical to keep in mind that AI should not be used to replace experts; rather, it should complement them. Because proper data is required to provide AI systems with access to the wide-ranging and diverse statistics of the population, allowing these systems to ultimately make accurate and realistic predictions using patient records could, from an ethical perspective, jeopardize patient confidentiality. Similar to different health care environments or education systems, the use of AI technology in pregnancy medicine has increased significantly and is increasingly employed in therapeutic and diagnostic medical decisions. Despite significant advancements in disease-related treatments and diagnosis prediction made by conventional statistical methods, AI stands to be the cornerstone and the most widely accepted strategy for the Fourth Industrial Revolution's massive wave of data-based advancements. Because maternal–fetal care involves keeping track of the wellbeing of two lives (mother and fetus), there is a greater need for stricter ethical standards when it comes to the deployment of AI technology. Thus, greater thought should be paid to the quantity, caliber, and correctness of information and contribution of research in this area.

References

[1] L. Davidson, M.R. Boland, Enabling pregnant women and their physicians to make informed medication decisions using artificial intelligence, J. Pharmacokinet. Pharmacodyn. 47 (2020) 305–318.

[2] R. Kaur, R. Kumar, M. Gupta, Deep neural network for food image classification and nutrient identification: a systematic review, Rev. Endocr. Metab. Disord. (2023) 1–21.

[3] P. Iftikhar, M.V. Kuijpers, A. Khayyat, A. Iftikhar, M.D. De Sa, Artificial intelligence: a new paradigm in obstetrics and gynecology research and clinical practice, Cureus 12 (2) (2020).

[4] A.M. Oprescu, G. Miró-Amarante, L. García-Díaz, V.E. Rey, A. Chimenea-Toscano, R. Martínez-Martínez, M.C. Romero-Ternero, Towards a data collection methodology for responsible artificial intelligence in health: a prospective and qualitative study in pregnancy, Inform. Fusion 83 (2022) 53–78.

[5] S. Polak, A. Mendyk, Artificial intelligence technology as a tool for initial GDM screening, Expert Syst. Appl. 26 (4) (2004) 455–460.

[6] R.O. Bahado-Singh, J. Sonek, D. McKenna, et al., Artificial intelligence and amniotic fluid multiomics: prediction of perinatal outcome in asymptomatic women with short cervix, Ultrasound Obstet. Gynecol. 54 (2019) 110–118.

[7] E.R. Norwitz, Artificial intelligence: can computers help solve the puzzle of parturition? Am. J. Obstet. Gynecol. 194 (5) (2006) 1510–1512.

[8] R.S. Guh, T.C.J. Wu, S.P. Weng, Integrating genetic algorithm and decision tree learning for assistance in predicting in vitro fertilization outcomes, Expert Syst. Appl. 38 (4) (2011) 4437–4449.

[9] A. Enshaei, C.N. Robson, R.J. Edmondson, Artificial intelligence systems as prognostic and predictive tools in ovarian cancer, Ann. Surg. Oncol. 22 (12) (2015) 3970–3975.

[10] L.K. Goodwin, S. Maher, Data mining for preterm birth prediction, Comput. Methods Prog. Biomed. 63 (1) (2000) 46–51.

[11] L.K. Goodwin, M.A. Iannacchione, W.E. Hammond, P. Crockett, S. Maher, K. Schlitz, Data mining methods find demographic predictors of preterm birth, Nurs. Res. 50 (6) (2001) 340–345.

[12] L.K. Goodwin, M.A. Iannacchione, Data mining methods for improving birth outcomes prediction, Outcomes Manag. 6 (2) (2002) 80–85.

[13] K.S. Lee, I.S. Song, E.S. Kim, K.H. Ahn, Determinants of spontaneous preterm labor and birth including gastroesophageal reflux disease and periodontitis, J. Korean Med. Sci. 35 (13) (2020) e105.

[14] K.S. Lee, K.H. Ahn, Artificial neural network analysis of spontaneous preterm labor and birth and its major determinants, J. Korean Med. Sci. 34 (24) (2019) e128.

[15] R. Fung, J. Villar, A. Dashti, L.C. Ismail, E. Staines-Urias, E.O. Ohuma, et al., Achieving accurate estimates of fetal gestational age and personalised predictions of fetal growth based on data from an international prospective cohort study: a population-based machine learning study, Lancet Digit Health 2 (7) (2020) e368–e375.

[16] K.S. Lee, H.Y. Kim, S.J. Lee, S.O. Kwon, S. Na, H.S. Hwang, et al., Prediction of newborn's body mass index using nationwide multicenter ultrasound data: a machine-learning study, BMC Pregnancy Childbirth 21 (1) (2021) 172.

[17] K.H. Ahn, K.S. Lee, Artificial intelligence in obstetrics, Obstet. Gynecol. Sci. 65 (2) (2022) 113–124.

[18] B. Cox, A.M. Sharma, R.A. Smith, S.B. Sharma, S.L. Ho, M.W. Gimenez-Badia, et al., Translational analysis of mouse and human placental protein and mRNA reveals distinct molecular pathologies in human preeclampsia, Mol. Cell. Proteomics 10 (12) (2011) M111.

[19] K.-J. Wang, A.M. Adrian, K.-H. Chen, K.-M. Wang, An improved electromagnetism-like mechanism algorithm and its application to the prediction of diabetes mellitus, J. Biomed. Inform. 54 (2015) 220–229.

[20] I. Jurisica, J. Mylopoulos, J. Glasgow, H. Shapiro, R.F. Casper, Case-based reasoning in IVF: prediction and knowledge mining, Artif. Intell. Med. 12 (1) (1998) 1–24.

[21] A. Mora-Sanchez, D.-I. Aguilar-Salvador, I. Nowak, Towards a gamete matching platform: using immunogenetics and artificial intelligence to predict recurrent miscarriage, NPJ Digit Med. 2 (1) (2019) 1–6.

[22] F.R. Jelovsek, D.R. Mattison, J.F. Young, Eliciting principles of hazard identification from experts, Teratology 42 (5) (1990) 521–533.

[23] K. Paydar, S.R. Niakan Kalhori, M. Akbarian, A. Sheikhtaheri, A clinical decision support system for prediction of pregnancy outcome in pregnant women with systemic lupus erythematosus, Int. J. Med. Inform. 97 (2016) 239–246.

[24] S. Yu, K.K. Tan, B.L. Sng, S. Li, A.T.H. Sia, Lumbar ultrasound image feature extraction and classification with support vector machine, Ultrasound Med. Biol. 41 (10) (2015) 2677–2689.

[25] H.-C. Lin, C.-T. Su, P.-C. Wang, An application of artificial immune recognition system for prediction of diabetes following gestational diabetes, J. Med. Syst. 35 (3) (2011) 283–289.

[26] J.A.A. Swart, R. Vos, T.F.J. Tromp, Interactive individualization: patient counselling and drug information supported by knowledge systems, Pharm. World Sci. 16 (3) (1994) 154–160.

Chapter 3

Early stage prediction of endometriosis cancer using fuzzy machine learning technique

Vijay Kumar[a] **and Kiran Pal**[b]

[a]*Manav Rachna International Institute of Research & Studies, Faridabad, Haryana, India,* [b]*Delhi Institute of Tool Engineering, Delhi, India*

3.1 Introduction

Real-life situations always involve deterministic and non-deterministic processes, where in non-deterministic process put challenges and the taken up by statistics with stability conditions. Real-life problems are often vague and beyond the control of precise notations of mathematics. For such vague situations, Zadeh generalized the concept of classical set theory and proposed the concept of fuzzy set theory, which has the potential to think like humans and represent incomplete information. Since the inception of fuzzy theory, research has intensified across domains of decision making. The generalization of fuzzy set theory was proposed by Atanassov [1] as intuitionistic fuzzy set (IFS) theory, which has both membership and non-membership grades. The complexity of the human body system may not be discussed through traditional approaches and the behavior of such systems has certain degree of fuzziness. Zadeh [2] proposed that fuzzy theory is a computational tool for handling incomplete and uncertain situations that can handle the problems of medical decision making effectively. Fuzzy theory accepts the information in linguistic ways and allows the system to describe the information via simple human-friendly rules. In any diagnostic process, doctors gather information about the patient from previous data, clinical findings, laboratory sample results, and other related sources of investigation. The knowledge gathered from these sources has a certain degree of incompleteness and certain things are overlooked or wrongly interpreted during investigation, resulting in incorrect treatment of the disease. Fuzzy models

Artificial Intelligence *and* Machine Learning
for Women's Health Issues
https://doi.org/10.1016/B978-0-443-21889-7.00014-2

model the problems of medical diagnosis and establish important symptoms and their patterns to propose the right treatment. These methods help in the configuration of treatments both initial as well as final diagnosis and form the basis for the diagnosis of diseases. Sanchez [3] established a fuzzy max-min relation between symptoms and diseases to represent the knowledge base and propose the right diagnosis. In this chapter, we discuss medical decision making of the diagnosis of early-stage endometrial cancer.

3.2   Multicriteria decision making

To provide computational support to decision makers for the evaluation of performance criteria to select the best course of action, multiple criteria decision-making (MCDM) techniques have been used. The generalization of Zadeh's fuzzy set [4] is IFS, which has applications in MCDM problems in special cases. Complex decision-making problems have been handled by the development of theory by Atanassov [5] and Bustince et al. [6]. Based on IFS, Kour et al. [7], Hong and Choi [8], Szmidt and Kacprzyk [9], and Atanassov et al. [10] proposed fuzzy-based algorithms used in MCDM problems. MCDM fuzzy models have been developed by several researchers such as Chen and Hwang [11], Kacprzyk et al. [12], Fodor and Runens [13], Herrera et al. [14], Bordogna et al. [15], Ma et al. [16], Saaty [17], and Hwang and Lin [18]. The prime application of the MCDM technique is to rank decision variables to achieve tangible goals. Also, MCDM models identify criteria sets to compare the alternatives and provide a robust solution to the problem at hand.

3.3   Aggregation

Aggregation is the process in which one representative value is assigned to the collected data, which could be some average, maximum, or minimum value. In many complex situations, different types of fuzzy and generalized fuzzy-based aggregation operators have been used extensively in different domains of research. Over the last couple of years, many fuzzy MCDM theories and methods have been used to develop various types of aggregation operators considering certain hypotheses, such as attributes and decision makers, at the same level of priority. In this chapter, we use priority operators for determining the suitable treatment for endometrial cancer.

3.4   Decision making

Decision making is a framework used to select the best alternative from a finite number of available alternatives. Decision making has received great interest from researchers across various disciplines, but in real life, the information available is ambiguous or imprecise. To solve problems of decision making with vague or imprecise information, fuzzy and IFS theories have emerged as powerful tools. In order to define the entropy function, fuzzy set and IFS theoretic approaches have been found useful in many real-life situations.

3.5   Medical diagnosis

Errors in disease diagnosis can lead to serious adverse impacts on a person, sometimes even leading to death. To reduce the risk of errors in disease diagnosis, many computational tools have been used to support healthcare practitioners and patients. Generally, diagnosis starts when a patient sees a doctor and the physician analyzes the patient's information to prescribe the suitable treatment on the basis of some knowledge base. The medical diagnosis process can be iterated in each stage and reconfigured or refined as per the need.

3.6   Some preliminaries

3.6.1   Intuitionistic fuzzy set (IFS)

Atanassov [1], suppose $X = \{x_1, x_2, \ldots, x_n\}$ is a discrete universe of discourse. An IFS T in X is given as: $T = \{<x, \theta_T(x), \phi_T(x)> | x \in X\}$ described by membership function $\theta_T(x) : X \to [0,1]$ and non-membership function $\phi_T(x) : X \to [0,1]$ of the element $x \in X$ where the function $\varphi_T(x) = 1 - \theta_T(x) - \phi_T(x)$ is defined as an intuitionistic index or hesitation index of x in T. In limiting case, if $\varphi_T(x) = 0$, IFS reduces automatically to fuzzy set.

3.6.2   Intuitionistic trapezoidal fuzzy number (ITrFN)

Wang [19] introduced the concept of ITrFN. Let T be an IFS in R, whose membership and non-membership functions are defined as follows:

$$\theta_T = \begin{cases} \dfrac{x - c_1}{c_2 - c_1}, & c_1 \leq x \leq c_2 \\ \theta_T, & c_2 \leq x \leq c_3 \\ \dfrac{c_4 - x}{c_4 - c_3}, & c_3 \leq x \leq c_4 \\ 0, & \text{otherwise} \end{cases} \quad \text{and} \quad \phi_T = \begin{cases} \dfrac{(c_2 - x) + \phi_T(x - c_{11})}{c_2 - c_{11}}, & c_{11} \leq x \leq c_2 \\ \phi_T, & c_2 \leq x \leq c_3 \\ \dfrac{(x - c_3) + \phi_T(c_{14} - x)}{c_{14} - c_3}, & c_3 \leq x \leq c_{14} \\ 0, & \text{otherwise} \end{cases}$$

$$(3.1)$$

respectively, where, $0 \leq \theta_T + \phi_T \leq 1$ and $c_1, c_2, c_3, c_4, c_{11}, c_{14} \in R$. Then T is called ITrFN. Denoting $T = \langle ([c_1, c_2, c_3, c_4]; \theta_T), ([c_{11}, c_2, c_3, c_{14}]; \phi_T) \rangle$. The shape of ITrFN originated from the fact that there are several points whose membership degree is 1.

Wang and Zhang [20] introduced operations performed on ITrFNs defined as:

Suppose, $T_1 = ([c_1, d_1, e_1, f_1]; \theta_{T_1}, \phi_{T_1})$ and $T_2 = ([c_2, d_2, e_2, f_2]; \theta_{T_2}, \phi_{T_2})$ are two ITrFNs, then

$$T_1 \oplus T_2 = \left([c_1 + c_2, d_1 + d_2, e_1 + e_2, f_1 + f_2]; \theta_{T_1} + \theta_{T_2} - \theta_{T_1}\theta_{T_2}, \phi_{T_1}\phi_{T_2} \right) \quad (3.2)$$

$$T_1 \otimes T_2 = \left([c_1 c_2, d_1 d_2, e_1 e_2, f_1 f_2]; \theta_{T_1}\theta_{T_2}, \phi_{T_1} + \phi_{T_2} - \phi_{T_1}\phi_{T_2} \right) \quad (3.3)$$

$$\alpha T_1 = \left([\alpha c_1, \alpha d_1, \alpha e_1, \alpha f_1]; 1 - (1 - \theta_{T_1})^\alpha, \phi_{T_1}{}^\alpha \right); \forall \alpha \geq 0 \quad (3.4)$$

$$T_1{}^\alpha = \left([c_1{}^\alpha, d_1{}^\alpha, e_1{}^\alpha, f_1{}^\alpha]; \theta_{T_1}{}^\alpha, 1 - \left(1 - \phi_{T_1}\right)^\alpha \right); \forall \alpha \geq 0 \quad (3.5)$$

3.6.3 Score and accuracy function of ITrFN

Wang and Zhang [21], score function measures the preciseness, whereas accuracy function ranks the set of real number $\mathfrak{R}$ and is defined as:

Let $T = ([c, d, e, f]; \theta_A, \phi_A)$ be an ITrFN, then

$$\Omega(T) = \Sigma(T) \times (\theta_A - \phi_A) \text{ is called the score function of } T \quad (3.6)$$

$$\Phi(A) = \Sigma(T) \times (\theta_A + \phi_A) \text{ is called the accuracy function of } T \quad (3.7)$$

where $\Sigma(A) = \frac{1}{8} \times [(c + d + e + f) \times (1 + \theta_A - \phi_A)]$ is called the expected value of A.

3.6.4 Ranking of ITrFN

Wang and Zhang [21] introduced some properties of ranking function.

If T_1 and T_2 are two random intuitionistic trapezoidal fuzzy numbers, then

$$\text{If } \Omega(T_1) > \Omega(T_2), \text{then } T_1 > T_2;$$

$$\text{If } \Omega(T_1) = \Omega(T_2) \text{ then}$$

$$\text{If } \Phi(T_1) = \Phi(T_2) \text{ and } T_1 = T_2$$

$$\text{If } \Phi(T_1) > \Phi(T_2) \text{ and } T_1 > T_2$$

The ranking principal on ITrFN is defined by using finite numbers of score and accuracy functions.

3.6.5 Intuitionistic trapezoidal fuzzy prioritized weighted average (ITrFPWA) operator

Let $T_i = ([c, d, e, f]; \theta, \phi)$ be the collection of ITrFNs.

$$\text{If } PWA(T_1, \ldots, T_n) = \frac{W_1}{\sum\limits_{J=1}^{n} W_j} T_1 \oplus \frac{W_2}{\sum\limits_{J=1}^{n} W_j} T_2 \oplus \ldots \oplus \frac{W_n}{\sum\limits_{J=1}^{n} W_j} T_n;$$

$$W_1 = 1 \text{ and } W_j = \prod_{k=1}^{j-1} \Omega(T_k)$$

$$\rightarrow PWA(T_1, \ldots, T_n) =$$

$$\left(\left[\left(\sum_{j=1}^{n} \frac{W_j c_j}{\sum\limits_{j=1}^{n} W_j} \right), \sum_{j=1}^{n} \left(\frac{W_j d_j}{\sum\limits_{j=1}^{n} W_j} \right), \sum_{j=1}^{n} \left(\frac{W_j e_j}{\sum\limits_{j=1}^{n} W_j} \right), \sum_{j=1}^{n} \left(\frac{W_j f_j}{\sum\limits_{j=1}^{n} W_j} \right) \right]; \right.$$

$$\left. 1 - \prod_{j=1}^{n} (1 - \theta_{A_j})^{\left(\frac{W_j}{\sum\limits_{j=1}^{n} W_j} \right)}, \prod_{k=1}^{l} (\phi_{A_j})^{\left(\frac{W_j}{\sum\limits_{j=1}^{n} W_j} \right)} \right)$$

$$(3.8)$$

3.7 Endometrial cancer types and treatments

Endometrial cancer is a gynecological malignancy often found in high-income countries. The incidence of the disease is increasing exponentially, but it is curable if diagnosed early. Endometrial cancer (also called endometrial carcinoma) starts when the growth of cells in the inner lining of the uterus, known as the endometrium (figure) is out of control. There are different types of endometrial cancer, including adenocarcinoma (endometrioid cancer), uterine carcinosarcoma, squamous cell carcinoma, small-cell carcinoma, transitional carcinoma, and serous carcinoma. There are subtypes and grades of these types of cancers, but we do not discuss them in this work.

In endometrial cancer, stage is an important factor for the selection of suitable treatment. Other factors include patient age, patient health status, type of cancer, and so on. The treatments for this type of cancer include surgery, radiation therapy, hormone therapy, targeted therapy, immunotherapy, and other treatments. Of these, surgery is the foremost treatment for almost all patients in which the uterus, fallopian tubes, and ovaries have to be removed and tested for malignant spread. Based on the stage of the disease, other treatments may be recommended. For pregnant patients, other treatments are recommended; surgery should wait until after the patient gives birth. For stage I patients with low grade tumors, surgery is the only treatment option. For high grade tumors, surgery followed by chemotherapy with or without radiation therapy may likely be recommended. In stage II patients, tumors have spread to the connective tissue of the cervix. In these cases, radiation therapy will be administered after surgery. In other cases, treatment may be reversed for lymph sampling. For high grade tumors, both radiation and chemotherapy may be administered after surgery. In stage III endometrial cancer, the cancer has spread outside the uterus. In these cases, radiation therapy is given before surgery, in hopes of shrinking the tumor. For advanced stage III cancers that cannot be treated with surgery, immune therapy will be recommended. Stage IV cancers have grown into the bladder or bowel and spread to the liver, lungs, or other organs. Radiation therapy and hormone therapy may be used to stop the spread of the tumor to other parts of the body. Sometimes, targeted drugs with or without immunotherapy may be administered to slow the growth of advanced stage cancers of this type.

Torre [22] Endometrial cancer is the fifth most common type of cancer, accounting for 4.8% of all cancers diagnosed in women. Sung [23], Endometrial cancer is the sixth most diagnosed gynecological disease with the greatest incidence in high-income countries. In 2020, there were an estimated 417,000 cases of and 97,000 deaths due to endometrial cancer. According to an American cancer report [24], there were approximately 60,050 newly diagnosed cases of and 10,470 deaths due to endometrial cancer in the United States in 2016, reflecting an increased incidence of 1%–2%. Due to the increase in population and lack of medical facilities, the number of patients with endometrial cancer is increasing exponentially; however, the number of available doctors is not sufficient to address these patients. Computational techniques are being used to help with medical diagnosis and enhance the efficiency of doctors. These techniques are decision-making tools that doctors can use to propose appropriate treatments for their patients. The process of decision making has uncertainties and the traditional techniques do not have the capabilities to deal with uncertain situations. Generalized fuzzy set theories are considered useful tools to prescribe suitable treatment for diseases. The problems of MCDM are interdisciplinary in nature and, when linked with generalized fuzzy sets, it adds strength to the concept. The expected values of ITrFN, as defined by Wang and Zhang [25] used fuzzy MCDM in uncertain situations. Guorong [26] presented some aggregation operators of intuitionistic trapezoidal fuzzy weighted arithmetic averaging (ITFWAA) over the expected values of ITrFN for the purpose of better decision making. Wan and Dong [27] defined the expected score of ITrFNs across the geometric point and presented operators of the trapezoidal intuitionistic fuzzy ordered weighted averaging operator and its hybridization. To streamline the concept of decision making, Yager [28–30] modeled the concept of prioritized operators. Wei [31] proposed the new version of prioritized aggregation operators defined by Yager [28,30] in a hesitant fuzzy environment. Yu and Xu [32] described the IF prioritization relationship of attributes and proposed prioritized operators of IF aggregation in MCDM, but the proposed operators have certain drawbacks. They cannot be used when the input is expressed in terms of ITrFNs and having difficulty in using with the problems of MADM in which attributes and decision makers are in different priority level. To overcome these shortcomings, Zhang [33] presented some prioritized operators containing ITrF information. These operators are very familiar with ITrFNs as input information and considering prioritization as the input arguments. Once the disease is diagnosed, it is very difficult for a doctor to develop reasoning for the specific treatment due to certain constraints. In this chapter, we consider endometrial cancer and aim to diagnose its suitable treatment computationally, using patient information. The objective of the study is to propose a support system specifically for medical decision making, which supports both doctors and patients in treatment and diagnosis. The proposed system is transparent and provides initial decision-making support to the decision makers after obtaining input information from three domain experts. Some suitable computational algorithms have been used to rank the diseases or treatments. To demonstrate the algorithm, we present a hypothetical case study.

3.8 Possible treatments for endometrial cancer

Fig. 3.1 presents possible treatments for endometrial cancer.

3.9 Relation between symptoms and treatments for endometrial cancer

Disease diagnosis is the art of determining a person's health using an available set of data. It is a challenging task. The process of medical diagnosis starts when a patient visits their doctor, and the doctor analyzes the situation by measuring certain parameters and symptoms. The diagnosis is then determined by considering the whole status of the patient. Then, a suitable treatment is prescribed and the whole process might be iterated with an aim to prevent and diagnose the disease. Therefore, there is a very close relationship between the diseases and the symptoms. In most cases, treatments can be clubbed to cure the disease. Let $T = \{T_1, T_2, ..., T_n\}$ and $S = \{s_1, s_2, ..., s_m\}$ be the finite set of treatments

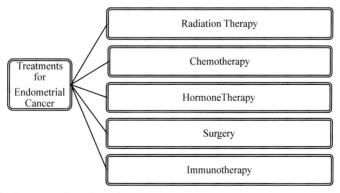

FIG. 3.1 Treatments for endometrial cancer.

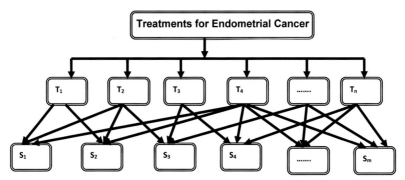

FIG. 3.2 Relationship between symptoms and their respective treatments for endometrial cancer.

and symptoms, respectively. Fig. 3.2 presents the relationship between treatments and symptoms of endometrial cancer.

3.10 Intuitionistic trapezoidal fuzzy prioritized weighted average (ITrFPWA) operators: An algorithm for the selection of suitable treatment of endometrial cancer

For the purpose of medical decision making, many researchers used the concept of MADM with the help of ITrFNs. To determine diseases, Durai and Iyengar [34] described rules and proposed a technique with the help of an algorithm. The approach for the diagnosis of heart diseases has been used by Bhatla and Jyoti [35]. Maryam [36] applied the approach proposed by Bhatla and Jyoti [35] to diagnose diabetes and asthma. In this chapter, we use the MADM approach over ITrF information for selecting the most appropriate treatment for endometrial cancer. Similarly, Yager [37] introduced ITrF prioritized average operators. Researchers such as Chen [38], Dehe [39], Delice [40], Dolan [41], Fishburn [42], Hung [43], Hwang [44], Kahraman [45], Kulak [46], and Xu [47] proposed various MCDM techniques for decision making. Chen [48] and Herrera [49] introduced operators in a fuzzy environment. In this chapter, the PWA operator has been used in fuzzy for the medical diagnosis of endometrial cancer.

In the following steps, ITrFPWA has been used and develop an algorithmic approach for ITrF information and is given as:.

Step 1. Transform the decision matrices $T^{(k)} = (c_{ij}^k)$ into the standardized decision matrices $R^{(k)}$.

Step 2. Calculate $W^p = (W_{ij}^{(p)})_{m \times n}; \forall p = 2, \dots q, \forall i = 1, \dots, m, \forall j = 1, 2, \dots, n$

(where) $W_{ij}^{(p)} = \prod_{k=1}^{p-1} \Omega\left(r_{ij}^{(k)}\right)$ and $T_{ij}^{(1)} = 1$

Step 3. Using the ITrFPWA operator (3.6.5).

The individual decision matrices $R^{(k)} = (r_{ij}^{(k)})_{m \times n}; \forall (k = 1, 2, \ldots, q)$ into the collective decision matrix $R = (r_{ij})_{m \times n} = (([e_{ij}, f_{ij}, g_{ij}, h_{ij}]; \theta_{ij}, \phi_{ij}))_{m \times n}$

$$r_{ij} = PWA\left(r_{ij}^{(1)}, \cdots, r_{ij}^{(q)}\right) = \frac{W_{ij}^{(1)}}{\sum_{p=1}^{q} W_{ij}^{(p)}} r_{ij}^{(1)} \oplus \cdots \oplus \frac{W_{ij}^{(q)}}{\sum_{p=1}^{q} W_{ij}^{(p)}} r_{ij}^{q}$$

$$= \left[\left[\sum_{k=1}^{q} \left(\frac{W_{ij}^{(k)} e_{ij}^{(k)}}{\sum_{p=1}^{q} W_{ij}^{(p)}} \right), \sum_{k=1}^{q} \left(\frac{W_{ij}^{(k)} f_{ij}^{(k)}}{\sum_{p=1}^{q} W_{ij}^{(p)}} \right), \sum_{k=1}^{q} \left(\frac{W_{ij}^{(k)} g_{ij}^{(k)}}{\sum_{p=1}^{q} W_{ij}^{(p)}} \right), \sum_{k=1}^{q} \left(\frac{W_{ij}^{(k)} h_{ij}^{(k)}}{\sum_{p=1}^{q} W_{ij}^{(p)}} \right) \right]; \right.$$

$$\left. 1 - \prod_{k=1}^{q} \left(1 - \theta_{ij}^{(k)}\right)^{\frac{W_{ij}^{(k)}}{\sum_{p=1}^{q} W_{ij}^{(p)}}}, \prod_{k=1}^{q} \left(\phi_{ij}^{(k)}\right)^{\frac{W_{ij}^{(k)}}{\sum_{p=1}^{q} W_{ij}^{(p)}}} \right)$$

(3.9)

Step 4. Calculate the matrix $[T]_{m \times n}$ based on the following equations:

$$W_{ij} = \prod_{k=1}^{j-1} \Omega(r_{ij}), T_{i1} = 1, \forall i = 1, \ldots, m, \forall j = 1, \ldots, n \qquad (3.10)$$

Step 5. Derive overall preference values r_i.

For each alternative x_i ($i = 1, 2, \ldots, m$), calculate collective overall preference r_i

$$r_{ij} = ITFPWA\left(r_{ij}^{(1)}, \ldots, r_{ij}^{(q)}\right) = \frac{W_{i1}^{(1)}}{\sum_{j=1}^{n} W_{ij}} r_{i1} \oplus \cdots \oplus \frac{W_{in}^{(n)}}{\sum_{j=1}^{n} W_{ij}} r_{in}$$

$$= \left[\left(\sum_{j=1}^{n} \left(\frac{W_{ij} e_{ij}}{\sum_{j=1}^{n} W_{ij}} \right), \sum_{j=1}^{n} \left(\frac{W_{ij} f_{ij}}{\sum_{j=1}^{n} W_{ij}} \right), \sum_{j=1}^{n} \left(\frac{W_{ij} g_{ij}}{\sum_{j=1}^{n} W_{ij}} \right), \sum_{j=1}^{n} \left(\frac{W_{ij} h_{ij}}{\sum_{j=1}^{n} W_{ij}} \right) \right) ; 1 - \prod_{j=1}^{n} (1 - \cdot_{ij})^{\frac{W_{ij}}{\sum_{j=1}^{n} W_{ij}}}, \prod_{j=1}^{n} (\phi_{ij})^{\frac{W_{ij}}{\sum_{j=1}^{n} W_{ij}}} \right] \qquad (3.11)$$

Step 6. Calculate the score and accuracy functions of r_i as follows:

$$\Omega(r_i) = \frac{1}{8} \times (e_i + f_i + g_i + h_i) \times (1 + \theta_i - \phi_i) \times (\theta_i - \phi_i) \qquad (3.12)$$

$$\Phi(r_i) = \frac{1}{8} \times (e_i + f_i + g_i + h_i) \times (1 + \theta_i - \phi_i) \times (\theta_i + \phi_i) \qquad (3.13)$$

Step 7. The best alternative can be selected by ranking all the alternatives x_i, $\forall i = 1, 2, 3, \ldots, m$.

Step 8. End.

All the information is given in the form of ITrFNs.

3.11 Evaluation of case study

In 2016, there were approximately 60,050 newly diagnosed cases of and 10,470 deaths due to endometrial cancer in the United States alone. A knowledge base has been framed with the help of domain experts as per the symptoms and the available treatments for the diseases. The most appropriate treatment for endometrial cancer can be selected by using the following algorithm.

Let $x_i (i = 1, 2, \ldots, 5)$ be the set of treatments available for endometrial cancer (see Fig. 3.2). The selection of the suitable treatment can be evaluated based on the symptoms $S = (s_1, \ldots, s_4)$ of the disease. The decision for the final treatment can be made after getting complete information on the disease and the whole status of the patient from the panel of decision makers as $D = (d_1, d_2, d_3)$. Tables 3.1–3.3 list the three ITrF decision matrices $T^{(k)} = (\alpha_{ij}^{(k)})_{m \times n} (k = 1, 2, 3)$.

TABLE 3.1 Intuitionistic trapezoidal fuzzy decision matrix provided by expert d_1.

Treatment/ Symptom	s_1	s_2	s_3	s_4
x_1	([2–4, 6]; 0.6, 0.3)	([4, 6, 8, 9]; 0.6, 0.3)	([2–4, 6]; 0.6, 0.4)	([3, 5, 6, 9]; 0.7, 0.1)
x_2	([1–3, 6]; 0.6, 0.4)	([6–9]; 0.8, 0.2)	([3, 5, 6, 8]; 0.8, 0.1)	([4, 6, 8, 9]; 0.7, 0.3)
x_3	([2–5]; 0.5, 0.3)	([5–7, 9]; 0.7, 0.3)	([2, 3, 5, 6]; 0.6, 0.4)	([3, 5, 6, 8]; 0.7, 0.1)
x_4	([1–3, 5]; 0.6, 0.4)	([6–9]; 0.8, 0.1)	([3, 4, 6, 7]; 0.8, 0.1)	([4, 6, 7, 9]; 0.6, 0.3)
x_5	([2–5]; 0.6, 0.3)	([5–8]; 0.6, 0.3)	([2, 3, 5, 9]; 0.6, 0.3)	([3, 4, 6, 8]; 0.7, 0.2)

TABLE 3.2 Intuitionistic trapezoidal fuzzy decision matrix provided by expert d_2.

Treatment/ Symptom	s_1	s_2	s_3	s_4
x_1	([5–7, 9]; 0.8, 0.1)	([2–4, 7]; 0.7, 0.3)	([3, 4, 6, 8]; 0.5, 0.4)	([5–7, 9]; 0.7, 0.3)
x_2	([1, 4, 5, 7]; 0.5, 0.4)	([2, 4–6]; 0.5, 0.4)	([2–4, 6]; 0.7, 0.3)	([2, 6–8]; 0.8, 0.1)
x_3	([5, 6, 8, 9]; 0.8, 0.1)	([2–5]; 0.6, 0.3)	([3–5, 8]; 0.5, 0.4)	([5–8]; 0.7, 0.2)
x_4	([1, 4, 5, 8]; 0.5, 0.4)	([2, 4, 5, 7]; 0.5, 0.4)	([2–5]; 0.6, 0.3)	([2, 5, 7, 8]; 0.8, 0.1)
x_5	([5–8]; 0.8, 0.2)	([2–4, 6]; 0.6, 0.3)	([3, 4, 6, 9]; 0.5, 0.3)	([5–7, 9]; 0.7, 0.1)

TABLE 3.3 Intuitionistic trapezoidal fuzzy decision matrix provided by expert d_3.

Treatment/ Symptom	s_1	s_2	s_3	s_4
x_1	([2, 3, 5, 7]; 0.5, 0.4)	([4, 5, 7, 8]; 0.5, 0.4)	([2, 3, 6, 8]; 0.6, 0.3)	([4–6, 8]; 0.8, 0.1)
x_2	([4, 6, 7, 9]; 0.6, 0.3)	([2, 3, 5, 6]; 0.7, 0.2)	([2, 3, 5, 6]; 0.7, 0.3)	([6–9]; 0.4, 0.6)
x_3	([2, 3, 5, 6]; 0.5, 0.4)	([4, 6–8]; 0.5, 0.4)	([2, 3, 6, 8]; 0.6, 0.4)	([4–7]; 0.8, 0.1)
x_4	([4, 6–8]; 0.6, 0.4)	([2–5]; 0.8, 0.2)	([2–4, 6]; 0.6, 0.3)	([6–9]; 0.3, 0.5)
x_5	([2, 4–6]; 0.5, 0.3)	([4, 6–8]; 0.6, 0.4)	([2, 5, 6, 8]; 0.6, 0.4)	([4–6, 8]; 0.8, 0.2)

3.11.1   Evaluation based on the given algorithm

Using step 1 and step 2 of the algorithm, the matrices $W^{(1)}$, $W^{(2)}$, and $W^{(3)}$ are as follows:

$$W^{(1)} = \begin{bmatrix} 1 & 1 & 1 & 1 \\ 1 & 1 & 1 & 1 \\ 1 & 1 & 1 & 1 \\ 1 & 1 & 1 & 1 \end{bmatrix} \quad W^{(2)} = \begin{bmatrix} 0.0488 & 0.1323 & 0.0300 & 0.2571 \\ 0.0300 & 0.3900 & 0.3347 & 0.2012 \\ 0.0257 & 0.1900 & 0.0343 & 0.2400 \\ 0.0263 & 0.4834 & 0.2975 & 0.1341 \\ 0.0418 & 0.1254 & 0.0766 & 0.1741 \end{bmatrix}$$

$$W^{(3)} = \begin{bmatrix} 0.0197 & 0.0106 & 0.0008 & 0.0489 \\ 0.0008 & 0.0100 & 0.0368 & 0.0813 \\ 0.0109 & 0.0079 & 0.0008 & 0.0579 \\ 0.0007 & 0.0133 & 0.0207 & 0.0513 \\ 0.0129 & 0.0061 & 0.0046 & 0.0567 \end{bmatrix}$$

Tables 3.4–3.6 present the decision matrices of the domain experts.

Table 3.7 presents the collective decision matrix using step 3 of the algorithm.

TABLE 3.4 Standardized decision matrix $R^{(1)}$ provided by expert d_1.

Treatment/ Symptom	s_1	s_2	s_3	s_4
x_1	([0, 0.142, 0.285, 0.571]; 0.6, 0.3)	([0.285, 0.571, 0.857, 1]; 0.6, 0.3)	([0, 0.142, 0.285, 0.571]; 0.6, 0.4)	([0.142, 0.428, 0.571, 1.0]; 0.7, 0.1)
x_2	([0, 0.125, 0.25, 0.625]; 0.6, 0.4)	([0.625, 0.75, 0.875, 1]; 0.8, 0.2)	([0.25, 0.50, 0.625, 0.875]; 0.8, 0.1)	([0.375, 0.625, 0.875, 1.00]; 0.7, 0.3)
x_3	([0, 0.142, 0.285, 0.42]; 0.5, 0.3)	([0.428, 0.571, 0.714, 1]; 0.7, 0.3)	([0, 0.142, 0.428, 0.571]; 0.6, 0.4)	([0.142, 0.428, 0.571, 0.857]; 0.7, 0.1)
x_4	([0, 0.125, 0.25, 0.50]; 0.6, 0.4)	([0.625, 0.75, 0.875, 1]; 0.8, 0.1)	([0.25, 0.375, 0.625, 0.75]; 0.8, 0.1)	([0.375, 0.625, 0.75, 1]; 0.6, 0.3)
x_5	([0, 0.142, 0.285, 0.428]; 0.6, 0.3)	([0.42, 0.57, 0.71, 0.85]; 0.6, 0.3)	([0, 0.142, 0.428, 1]; 0.6, 0.3)	([0.142, 0.285, 0.571, 0.857]; 0.7, 0.2)

TABLE 3.5 Standardized decision matrix $R^{(2)}$ provided by expert d_2.

Treatment/ Symptom	s_1	s_2	s_3	s_4
x_1	([0.4286, 0.5714, 0.7143, 1.00]; 0.8, 0.1)	([0, 0.1429, 0.2857, 0.7143]; 0.7, 0.3)	([0.142, 0.285, 0.571, 0.857]; 0.5, 0.4)	([0.428, 0.571, 0.714, 1]; 0.7, 0.3)
x_2	([0, 0.4286, 0.5714, 0.8571]; 0.5, 0.4)	([0.142, 0.428, 0.571, 0.714]; 0.5, 0.4)	([0.142, 0.285, 0.428, 0.714]; 0.7, 0.3)	([0.142, 0.714, 0.857, 1; 0.8, 0.1)
x_3	([0.4286, 0.5714, 0.8571, 1.00];0.8, 0.1)	([0, 0.142, 0.285, 0.428];0.6, 0.3)	([0.142, 0.285, 0.428, 0.857]; 0.5, 0.4)	([0.428, 0.571, 0.714, 0.857]; 7,0.2)
x_4	([0, 0.4286, 0.5714, 1.00]; 0.5, 0.4)	([0.142, 0.428, 0.571, 0.857]; 0.5, 0.4)	([0.142, 0.285, 0.428, 0.571]; 0.6, 0.3)	([0.1429, 0.5714, 0.8571, 1]; 0.8, 0.1)
x_5	([0.4286, 0.5714, 0.7143, 0.8571];0.8, 0.2)	([0, 0.142, 0.285, 0.571]; 0.6, 0.3)	([0.142, 0.285, 0.571, 1]; 0.5, 0.3)	([0.4286, 0.5714, 0.7143, 1]; 0.7, 0.1)

TABLE 3.6 Standardized decision matrix $R^{(3)}$ provided by expert d_3.

Treatment/ Symptom	s_1	s_2	s_3	s_4
x_1	([0, 0.166, 0.50, 0.83]; 0.5, 0.4)	([0.33, 0.50, 0.83, 1]; 0.5, 0.4)	([0, 0.166, 0.6667, 1]; 0.6, 0.3)	([0.33, 0.50, 0.666, 1.00]; 0.8, 0.1)
x_2	([0.285, 0.571, 0.714, 1]; 0.6, 0.3)	([0, 0.142, 0.428, 0.571]; 0.7, 0.2)	([0, 0.142, 0.428, 0.571]; 0.7, 0.3)	([0.571, 0.714, 0.857, 1.00]; 0.4, 0.6)
x_3	([0, 0.166, 0.5, 0.666]; 0.5, 0.4)	([0.33, 0.666, 0.83, 1.0]; 0.5, 0.4)	([0, 0.166, 0.666, 1.00]; 0.6, 0.4)	([0.33, 0.50, 0.666, 0.83]; 0.8, 0.1)

TABLE 3.6 Standardized decision matrix $R^{(3)}$ provided by expert d_3—cont'd

Treatment/ Symptom	s_1	s_2	s_3	s_4
x_4	([0.28, 0.57, 0.71, 0.85]; 0.6, 0.4)	([0, 0.142, 0.285, 0.428]; 0.8, 0.2)	([0, 0.142, 0.285, 0.571]; 0.6, 0.3)	([0.571, 0.714, 0.857, 1.00]; 0.3, 0.5)
x_5	([0, 0.33, 0.50, 0.666]; 0.5, 0.3)	([0.33, 0.666, 0.83,1.00]; 0.6, 0.4)	([0,0.50, 0.666, 1.0]; 0.6, 0.4)	([0.33, 0.50, 0.666, 1.0]; 0.8, 0.2)

TABLE 3.7 Collective decision matrix R.

Treatment/ Symptom	s_1	s_2	s_3	s_4
x_1	([0.019, 0.162, 0.309, 0.595]; 0.6109, 0.2868)	([0.253, 0.521, 0.790, 0.966]; 0.612, 0.300)	([0.0042, 0.1470, 0.2943, 0.5801]; 0.5974, 0.3999)	([0.2062, 0.4594, 0.6031, 1.00]; 0.7045, 0.1241)
x_2	([0.0002, 0.1342, 0.2597, 0.6320]; 0.5974, 0.399)	([0.4862, 0.6561, 0.7873, 0.9174]; 0.7411, 0.2426)	([0.2171, 0.4381, 0.5718, 0.8276]; 0.7768, 0.1347)	([0.3510, 0.6447, 0.8711, 1.0]; 0.7059, 0.2638)
x_3	([0.0106, 0.1537, 0.3021, 0.4453]; 0.5112, 0.2928)	([0.3600, 0.5041, 0.6471, 0.9094]; 0.6849, 0.3006)	([0.0047, 0.1476, 0.4288, 0.5812]; 0.5970, 0.40)	([0.2042, 0.4582, 0.6021, 0.8561]; 0.7054, 0.1137)
x_4	([0.0002, 0.1331, 0.2585, 0.5130]; 0.5977, 0.40)	([0.4637, 0.6408, 0.7717, 0.9488]; 0.7311, 0.1574)	([0.2219, 0.3512, 0.5753, 0.7069]; 0.7636, 0.1304)	([0.3572, 0.6228, 0.7668, 1.00]; 0.6211, 0.2709)
x_5	([0.0170, 0.1622, 0.3053, 0.4485]; 0.6098, 0.2952)	([0.3806, 0.5245, 0.6674, 0.8263]; 0.60, 0.3005)	([0.0101, 0.1545, 0.4397, 1.00]; 0.5936, 0.3004)	([0.1920, 0.3360, 0.5960, 0.8839]; 0.7056, 0.1813)

Using step 4, the following matrix is:

$$W = \begin{bmatrix} 1 & 0.0583 & 0.0075 & 0.0002 \\ 1 & 0.0303 & 0.0081 & 0.0022 \\ 1 & 0.0303 & 0.0049 & 0.0002 \\ 1 & 0.0268 & 0.0085 & 0.0020 \\ 1 & 0.0482 & 0.0056 & 0.0004 \end{bmatrix}$$

Using step 5, the overall preference value r_i against the alternatives x_i; $(i = 1, 2, \ldots, 5)$ follows as:

$$r_1 = ([0.0323, \ 0.1824, \ 0.3355, \ 0.6161]; 0.6109, \ 0.2882)$$
$$r_2 = ([0.0168, \ 0.1528, \ 0.2788, \ 0.6426]; 0.6046, \ 0.3905)$$
$$r_3 = ([0.0209, \ 0.1640, \ 0.3129, \ 0.4596]; 0.5180, \ 0.2934)$$
$$r_4 = [0.0147, \ 0.1489, \ 0.2754, \ 0.5268]; 0.6037, \ 0.3866)$$
$$r_5 = ([0.0336, \ 0.1788, \ 0.3227, \ 0.4689]; 0.6093, \ 0.2954)$$

Using step 6, score and accuracy functions of r_i are calculated as:

$$\Omega(r_1) = 0.0622; \Omega(r_2) = 0.0355; \Omega(r_3) = 0.0329; \Omega(r_4) = 0.0319;$$
$$\Omega(r_5) = 0.0518$$

Finally, using step 7, the ranking of the available alternatives x_i is given as:

$$\Omega(r_1) \succ \Omega(r_5) \succ \Omega(r_2) \succ \Omega(r_3) \succ \Omega(r_4)$$

Table 3.8 lists the values of score function of each treatment for endometrial cancer.

3.12 Result and discussion

The score function is calculated using the ITFPWA operator (see Table 3.8). The results are graphically shown in Fig. 3.3. The highest score value is for radiation therapy. Therefore, the best treatment for this case is radiation therapy.

TABLE 3.8 Score function of various treatments of endometrial cancer.

Treatment	Score function
Radiation therapy	0.0662
Chemotherapy	0.0355
Hormone therapy	0.0329
Surgery	0.0319
Immunotherapy	0.0518

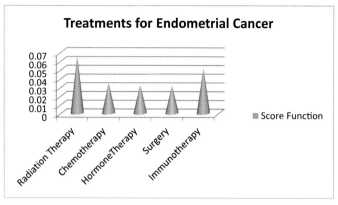

FIG. 3.3   Score function for various endometrial cancer treatments.

Since the optimal decision is not always the robust decision, we proceed further to calculate the score and accuracy function of the overall preference value and observed that $\Omega(r_1)$ is again selected as the optimal decision due to having the highest score value among the available score values. Therefore, the best treatment proposed using the given information is radiation therapy.

3.13   Conclusion

In this chapter we proposed an MADM approach for selecting the most suitable treatment for endometrial cancer in the context of ITrFPWA operator to aggregate ITrFNs for each of the alternatives. Ranking was done and the optimal decision evaluated based on score and accuracy values. The main objective of the study is to determine the most suitable treatment for endometrial cancer. We ranked all the treatment options with the help of the given operator and the ITrF information provided by the domain experts. Our proposed method can be embedded with other problems of medical decision making to obtain better decisions based on the clinical diagnosis of diseases. Also, the proposed method may help decision makers handle other real-life decision-making problems.

References

[1] K. Atanassov, Intuitionistic fuzzy sets, Fuzzy Sets Syst. 20 (1986) 87–96.
[2] L.A. Zadeh, Bio. appl. of the theory of fuzzy sets and sys, in: The Proc. of an Intel. Sym. on Biocyb. of Central Ner. Sys., Little, Brown and Co, 1969, pp. 199–206.
[3] E. Sanchez, Medical Diagnosis and Composite Fuzzy Relations, Fuzzy Set Theory and Applications, 1979, pp. 437–444.
[4] L.A. Zadeh, Fuzzy sets, Inf. Control 8 (1965) 338–356.
[5] K. Atanassov, Int. Fuzzy Sets, Theory and Applications, Physica-Verlag, 1999.
[6] H. Bustince, F. Herrera, J. Montero, Fuzzy Sets and their Extensions Representation, Aggregation and Models, Physica-Verlag, 2007.

[7] S. Kour, R. Kumar, M. Gupta, J.M. Tan, Study on detection of breast cancer using machine learning, in: 2021 International Conference in Advances in Power, Signal, and Information Technology, IEEE, 2021, pp. 1–9.

[8] D.H. Hong, C.H. Choi, Multicriteria fuzzy decision-making problems based on vague set theory, Fuzzy Sets Syst. 114 (2000) 103–113.

[9] E. Szmidt, J. Kacprzyk, Distances between intuitionistic fuzzy sets, Fuzzy Sets Syst. 114 (2000) 505–518.

[10] K. Atanassov, G. Pasi, R.R. Yager, Intuitionistic fuzzy interpretations of multi-criteria multi-person and multi-measurement tool decision making, Int. J. Syst. Sci. 36 (2005) 859–868.

[11] S.J. Chen, C.L. Hwang, Fuzzy Multiple Attribute Decision Making, Springer-Verlag, 1992.

[12] J. Kacprzyk, M. Fedrizzi, H. Nurmi, Group decision making and consensus under fuzzy preferences and fuzzy majority, Fuzzy Sets Syst. 49 (1992) 21–31.

[13] J.C. Fodor, M. Rubens, Fuzzy Preference Modelling and Multicriteria Decision Support, Kluwer Academic Publisher, 1994.

[14] F. Herrera, E. Herrera, Viedma, J.L. Verdegay, A linguistic decision process in group decision making, Group Decis. Negot. 5 (1996) 165–176.

[15] G. Bordogna, M. Fedrizzi, G. Pasi, A linguistic modeling of consensus in group decision making based on OWA operators, IEEE Trans. Syst. Man Cybernet. 27 (1) (1997) 126–132.

[16] J. Ma, Z.P. Fan, L.H. Huang, A subjective and objective integrated approach to determine attribute weights, Eur. J. Oper. Res. 112 (1999) 397–404.

[17] T.L. Saaty, The Analytic Hierarchy Process, McGraw-Hill, 1980.

[18] C.L. Hwang, M.J. Lin, Group decision making under multiple criteria: methods and applications, in: Lecture Notes in Economics and Mathematical Systems, Springer-Verlag, 1987.

[19] J.Q. Wang, Overview on fuzzy multi-criteria decision-making approach, Control Decis. 23 (2008) 601–606.

[20] J.Q. Wang, Z.H. Zhang, A method of multi-criteria decision-making with incomplete certain information based on intuitionistic trapezoidal fuzzy number, Control Decis. 24 (2009) 226–230.

[21] J.Q. Wang, Z.H. Zhang, Aggregation operators on intuitionistic trapezoidal fuzzy number and its application to multi-criteria decision making problems, J. Syst. Eng. Electron. 20 (2009) 321–326.

[22] L.A. Torre, F. Bray, R.L. Siegel, J. Ferlay, J. Lortet-Tieulent, A. Jemal, Global Cancer Statistics, 2012.

[23] H. Sung, J. Ferlay, R.L. Siegel, M. Laversanne, I. Soerjomataram, A. Jemal, Global Cancer Statistics 2020: GLOBOCAN Est. of incidence and mortality worldwide for 36 cancers in 185 countries, CA Cancer J. Clin. 71 (3) (2021) 209–249.

[24] American Cancer Society, Cancer Facts & Figures 2016, 2016, pp. 1–9.

[25] J.Q. Wang, Z.H. Zhang, Multi-criteria decision-making method with incomplete certain information based on intuitionistic fuzzy number, Control Decis. 23 (2008) 1145.

[26] X. Guorong, Models for multiple attribute decision making with intuitionistic trapezoidal information, Int. J. Adv. Comp. Tech. 3 (6) (2011) 21–25.

[27] S.P. Wan, J.Y. Dong, Method of trapezoidal intuitionistic fuzzy number for multi-attribute group decision, Control Decis. 25 (5) (2010) 773–776.

[28] R.R. Yager, OWA aggregation over a continuous interval argument with applications to decision making, IEEE Trans. Syst. Man Cybernet. 34 (2004) 1952–1963.

[29] R.R. Yager, Prioritized aggregation operators, Int. J. Behav. Sci. 16 (6) (2008) 538–544.

[30] R.R. Yager, Prioritized OWA aggregation, Fuzzy Optim. Decis. Making 8 (2009) 245–262.

[31] G.W. Wei, Hesitant fuzzy prioritized operators and their application to multiple attribute decision making, Knowl.-Based Syst. 31 (2012) 176–182.

[32] X. Yu, Z. Xu, Prioritized intuitionistic fuzzy aggregation operators, Inf. Fusion 14 (2013) 108–116.

[33] Z. Zhang, Fuzzy prioritized operators and their application to multiple attribute group decision making, Br. J. Maths Comput. Sci. (2014) 1951–1998.

[34] M.S. Durai, N. Iyengar, Effective analysis and diagnosis of lung cancer using fuzzy rules, Int. J. Eng. Sci. Technol. 2 (6) (2010) 2102–2108.

[35] N. Bhatla, K. Jyoti, A novel approach for heart disease diagnosis using data mining and fuzzy logic, Int. J. Comput. Appl. 54 (17) (2012) 16–21.

[36] M. Zolnoori, M.H.F. Zarandi, M. Moin, M. Taherian, Fuzzy rule-based expert system for evaluating level of asthma control, J. Infect. Dev. Countr. 171–184 (2010).

[37] R.R. Yager, Modeling prioritized multi-criteria decision making, IEEE Trans. Syst. Man Cybernet. 34 (2004) 2396–2404.

[38] S.M. Chen, J.M. Tan, Handling multicriteria fuzzy decision-making problems based on vague set theory, Fuzzy Sets Syst. 67 (1994) 163–172.

[39] B. Dehe, D. Bamford, Development, test and comparison of two multiple criteria decision analysis (MCDA) models: a case of healthcare infrastructure location, Expert Syst. Appl. 42 (2015) 6717–6727.

[40] E.K. Delice, S. Zegerek, Ranking occupational risk levels of emergency departments using a new fuzzy MCDM model: a case study in Turkey, Appl. Math. Inf. Sci. 10 (6) (2016) 2345–2356.

[41] J. Dolan, Multi-criteria clinical decision support: a primer on the use of multiple-criteria decision-making methods to promote evidence-based, patient-centered healthcare, Patient: Patient-Centered Outcomes Res. 3 (2010) 229–248.

[42] P.C. Fishburn, A comparative analysis of group decision methods: a comparative analysis of group decision methods, Behav. Sci. 16 (6) (1971) 538–544.

[43] C.C. Hung, L.H. Chen, A multiple criteria group decision making model with entropy weight in an intuitionistic fuzzy environment, in: X. Huang, S.I. Ao, O. Castillo (Eds.), Intelligent Automation and Computer Engineering, Lecture Notes in Electrical Engineering, Springer, Dordrecht, 2009.

[44] C.L. Hwang, K. Yoon, Multiple Objective Decision Making Methods and Applications: a State-of-the-Art Survey, Lecture Notes in Economics and Mathematical Systems, Springer-Verlag, New York, 1981.

[45] C. Kahraman, S.C. Onar, B. Oztaysi, Fuzzy multicriteria decision-making: a literature review, International Journal of Computational Intelligence Systems 8 (4) (2015) 637–666.

[46] O. Kulak, H.G. Goren, A.A. Supciller, A new multi criteria decision making approach for medical imaging systems considering risk factors, Appl. Soft Comput. 35 (2015) 931–941.

[47] Z.S. Xu, H. Hu, Projection models for intuitionistic fuzzy multiple attribute decision making, Int. J. Inf. Technol. Decis. Mak. 9 (2010) 267–280.

[48] T. Chen, C. Li, Determining objective weights with intuitionistic fuzzy entropy measures: a comparative analysis, Inf. Sci. 180 (2010) 4207–4222.

[49] F. Herrera, E. Viedma, On the linguistic OWA operator and extensions, in: R.R. Yager, J. Kacprzyk (Eds.), The Ordered Weighted Averaging Operators, Springer, Boston, MA, 1997.

Chapter 4

Artificial intelligence approaches for ultrasound examination in pregnancy

Somya Srivastava[a], Disha Mohini Pathak[a], and Gaurav Dubey[b]
[a]ABES Engineering College, Ghaziabad, India, [b]KIET Group of Institution, Ghaziabad, India

4.1 Introduction

Ultrasound examination is a valuable tool used in obstetrics and gynecology to monitor fetal development and diagnose potential fetal anomalies. However, the interpretation of ultrasound images is subject to human error and can be influenced by the experience and expertise of the clinician performing the exam. Artificial intelligence (AI) has emerged as a promising solution to these challenges, offering the potential to improve the efficiency of ultrasound examination. In this chapter, we explore the contributions of several authors to the field of AI in ultrasound examination in pregnancy [1]. One of the early pioneers in AI applications for ultrasound was R. Nigel Hughson, who authored the paper "Intelligent Processing and Analysis of Ultrasound Images" in the *Journal of Medical Engineering & Technology*. Hughson proposed an AI-based approach for ultrasound image analysis, which he called "intelligent image processing" (IIP). He suggested that IIP could be used to automate the process of identifying regions of interest in ultrasound images, such as the fetal head or heart, and measuring key parameters for diagnosis [2]. A team of researchers from the University of Melbourne, Australia, led by Dr. Christos Bergeles published "Real-time ultrasound segmentation using level sets and CUDA" in the *Journal of Real-Time Image Processing*. This study proposed an AI-based approach for real-time segmentation of ultrasound images [3] using level sets and CUDA parallel computing. The authors demonstrated that their method could accurately segment fetal structures in real time, with potential applications in fetal biometry and anomaly detection [4]. In more recent years, deep learning (DL) approaches have gained popularity in the field of AI-based ultrasound image analysis. The authors of "Fetal Ultrasound Image Segmentation

Artificial Intelligence *and* Machine Learning
for Women's Health Issues
https://doi.org/10.1016/B978-0-443-21889-7.00012-9

with Deep Learning: A Review," provided an overview of the use of DL techniques, such as convolutional neural networks (CNNs), for fetal ultrasound image segmentation. They discussed the benefits and limitations of these methods and presented several case studies demonstrating their potential for accurate and efficient diagnosis of fetal anomalies [5]. Overall, the contributions of these authors and others have advanced the development of AI approaches for ultrasound examination in pregnancy. These technologies offer the potential to improve the accuracy and efficiency of prenatal diagnosis and improve outcomes for both mothers and fetus. Ultrasound technology has revolutionized obstetrics by providing a noninvasive method to monitor fetal development during pregnancy. As ultrasound technology has improved over the years, so too has examination and analysis of images of the developing fetus. However, interpreting these images can be complex and time-consuming, requiring extensive training and experience on the part of radiologists. This is where machine learning (ML) algorithms come in, providing a powerful tool to assist in image analysis and interpretation.

In recent years, ultrasonic testing has made growing use of ML and DL algorithms to aid in image processing and interpretation. In order to learn from data and make decisions or predictions, these algorithms use computational methods. ML algorithms can assist radiologists in the analysis of obstetric ultrasound images and the detection of probable anomalies or regions of concern with increased speed and accuracy. Segmentation of ultrasound images is a crucial use of ML techniques in obstetric ultrasound. Fig. 4.1 illustrates the process of image segmentation, which entails isolating target regions or structures from the image's backdrop. In obstetric ultrasound, this is especially helpful in determining the location of the fetus, the placenta, and the amniotic fluid. A CNN is just one type of ML method that can be used for image segmentation. CNNs are examples of DL algorithms that can be taught to identify

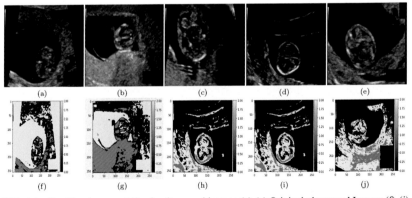

FIG. 4.1 Fetal head segmentation in ultrasound images (a)-(e) Original ultrasound Images (f)-(j) Segmentation results using K-Means.

visual patterns. A CNN may be trained on a large collection of ultrasound pictures, allowing the algorithm to properly identify and separate the region of interest in new images. ML methods can then be used to extract useful features from the segmented and labeled image data. For example, algorithms can assess the volume of amniotic fluid or calculate the size of the fetal head automatically. These characteristics can subsequently be used in screening for anomalies or monitoring fetal growth. Classification and diagnosis in obstetric ultrasound using ML techniques is also possible. Ultrasound images, for instance, can be categorized by the presence or absence of anomalies, such as congenital heart problems, using a trained algorithm. This can aid in the detection of prospective issues and the radiologists' ability to make precise diagnosis. ML algorithms of various kinds are widely employed in obstetric ultrasound. While random forests (RFs) can be used for both classification and feature extraction, support vector machines (SVMs) are more commonly utilized for classification. Ultrasound time-series analysis is often performed using recurrent neural networks (RNNs), whereas deep belief networks (DBNs) are employed for image reconstruction.

There are various obstacles that must be overcome before ML algorithms can truly revolutionize obstetric ultrasonography. For instance, there is a dearth of high-quality big data sets, which cannot substitute human intelligence. Radiologists will still need to use their clinical expertise and judgment when interpreting ultrasound pictures.

Overall, it is safe to say that obstetric ultrasound is becoming increasingly dependent on ML algorithms to aid radiologists in picture processing and interpretation. Better results for mothers and their growing fetuses are possible when these techniques become more widely used in clinical practice as technology advances. Nevertheless, radiologists will still play a significant role in the interpretation and analysis of ultrasound pictures, as ML algorithms cannot replace human skill and judgment. In the next section, we discuss the use of ML and DL models in obstetric ultrasound. In Section 4.3, we examine applications of AI in obstetric ultrasound.

4.2 Machine learning and deep learning models used in ultrasound examination in pregnancy

There are several ML models that have been used for ultrasound examination in pregnancy, including linear regression, which can be used for predicting fetal weight and estimating gestational age based on fetal biometry measurements. RFs and SVMs can be used for fetal biometry measurements and fetal weight estimation, as can k-nearest neighbors and CNNs. A Gaussian mixture model can be used for fetal biometry measurements. Table 4.1 presents the details of these models and some others.

TABLE 4.1 Machine learning and deep learning models used in ultrasound pregnancy examination.

Methods	Application in ultrasound pregnancy examination
Support vector machines (SVMs)	SVMs are used for classifying ultrasound images into different classes based on fetal health or abnormalities. They can be employed to identify specific conditions such as congenital heart defects, placenta previa, or gestational age estimation
Convolutional neural networks (CNNs)	CNNs are often used for image recognition tasks in ultrasound pregnancy examinations. They can detect and classify various fetal structures, such as the heart, brain, and limbs, as well as evaluate fetal growth and development. CNNs can also be employed in anomaly detection and segmentation of the ultrasound images
Random forests (RFs)	For both classification and regression problems, many researchers turn to RF, a prominent ensemble learning method. In pregnancy examination, RF can classify ultrasound images based on fetal health, predict gestational age, and identify various fetal conditions, such as intrauterine growth restriction (IUGR) or chromosomal abnormalities
Long short-term memory (LSTM)	LSTMs are a type of recurrent neural network (RNN) that can handle sequences of data. In pregnancy examination, LSTMs can be applied to time-series data from Doppler ultrasound for monitoring fetal heart rate, blood flow patterns, and assessing the risk of complications such as preeclampsia or preterm labor
K-nearest neighbors (KNN)	KNN is a simple and effective ML model for classification tasks. It can be employed in ultrasound pregnancy examination to classify images based on fetal health and developmental stages or to detect specific abnormalities, such as neural tube defects or skeletal dysplasia, by comparing the ultrasound images to a labeled dataset of similar images
Decision trees (DTs)	DTs can be used for both classification and regression tasks in ultrasound pregnancy examination. They can help identify various fetal conditions, predict gestational age, and classify images based on developmental stages or health status. DTs are also useful for feature selection and understanding the importance of different attributes
Principal component analysis (PCA)	PCA is a dimensionality reduction technique that can be used in ultrasound pregnancy examination to compress high-dimensional ultrasound images into a lower-dimensional representation. This can help reduce noise, enhance image quality, and improve the performance of other ML models when analyzing the images for classification, segmentation, or anomaly detection

TABLE 4.1 Machine learning and deep learning models used in ultrasound pregnancy examination—cont'd

Methods	Application in ultrasound pregnancy examination
Deep learning autoencoders	Autoencoders are unsupervised DL models that can be applied to denoising and feature extraction tasks in ultrasound pregnancy examination. They can help in improving image quality, detecting and segmenting fetal structures, and extracting relevant features from the images that can be fed into other ML models for classification or regression tasks
Gradient boosting machines (GBM)	GBM is an ensemble learning method used for classification and regression tasks. In pregnancy examination, GBM can be employed to classify ultrasound images based on fetal health, predict gestational age, and identify various fetal conditions, such as IUGR or chromosomal abnormalities. GBM is known for its accuracy and ability to handle large and complex datasets
Bayesian networks (BNs)	BNs are probabilistic graphical models that can be used for inference, prediction, and decision making in ultrasound pregnancy examination. BNs can help estimate the probability of specific fetal conditions, abnormalities, or outcomes based on the observed ultrasound data, as well as incorporate expert knowledge and prior information into the analysis

4.3   Applications of artificial intelligence in ultrasound examination during pregnancy

AI is finding more and more applications in the medical field, especially in gynecology and obstetrics. AI techniques are being applied to the field of obstetric ultrasound in order to aid doctors in the detection and treatment of fetal anomalies and problems. Clinically useful data, such as fetal weight, brain structure, and heart function, can be extracted from ultrasound pictures using these methods thanks to the application of DL algorithms. The use of AI in ultrasound testing has the potential to enhance its precision and efficiency, leading to better health outcomes for patients.

- **Fetal biometry**: AI algorithms are used for accurate measurement of fetal biometric parameters, such as head circumference, abdominal circumference, and femur length.
- **Fetal weight estimation**: AI algorithms are used to estimate fetal weight accurately, which is essential for managing high-risk pregnancies.

- **Fetal anomaly detection**: AI algorithms can assist in the detection of fetal anomalies, such as congenital heart defects, brain anomalies, and abdominal wall defects.
- **Fetal brain imaging**: AI algorithms can help in the segmentation and analysis of fetal brain structures and detecting abnormalities, such as ventriculomegaly.
- **Placental assessment**: AI algorithms can assist in the assessment of placental function, such as blood flow and thickness, which can help in the diagnosis of conditions like preeclampsia.

4.3.1 Fetal biometry

Several studies have investigated the feasibility of using DL and ML methods for automated fetal biometric monitoring from ultrasound images. This research proves that these methods have the potential to enhance the precision and efficiency of fetal biometry measurement. Automated fetal biometric measurement using ultrasound pictures has been proposed using a variety of DL and ML models [6–13]. These devices were both very accurate and a convenient alternative to time-consuming manual measuring. Researchers in Ref. [14], using a deep CNN to estimate fetal biometry, found their model to be more accurate than previous approaches. The results of this research show great promise for enhancing prenatal care and decreasing the likelihood of unfavorable pregnancy outcomes through the provision of precise and timesaving fetal biometry assessment. Table 4.2 presents a selection of current publications on fetal biometry.

TABLE 4.2 Summary of deep learning and machine learning techniques for automated fetal biometric measurement from ultrasound images: Reported performance metrics and dataset size.

Reference	Year	Deep learning/ machine learning technique	Performance metrics	Dataset size
Salem et al. [6]	2022	U-Net	Dice coefficient (0.947), mean absolute error (0.51 mm)	265 ultrasound images
Ayyildiz and Yildirim [7]	2021	Convolutional neural network (CNN)	Accuracy (96.28%), mean absolute error (1.20 mm)	1452 ultrasound images

TABLE 4.2 Summary of deep learning and machine learning techniques for automated fetal biometric measurement from ultrasound images: Reported performance metrics and dataset size—cont'd

Reference	Year	Deep learning/ machine learning technique	Performance metrics	Dataset size
Mahmood et al. [8]	2021	Deep convolutional neural network (DCNN)	Accuracy (96.0%), SEN (92.0%), SPE (98.0%)	2274 ultrasound images
Wang et al. [9]	2021	Convolutional neural network (CNN)	Accuracy (97.73%), mean absolute error (0.66 mm)	702 ultrasound images
Chen et al. [10]	2021	Convolutional neural network (CNN)	Accuracy (94.3%), mean absolute error (1.13 mm)	404 ultrasound images
Lee et al. [11]	2021	Multitask learning network	Accuracy (95.9%), mean absolute error (1.03 mm)	1140 ultrasound images
Wang et al. [12]	2021	Convolutional neural network (CNN)	Accuracy (96.4%), mean absolute error (1.15 mm)	471 ultrasound images
Huang et al. [13]	2020	Random forest (RF), support vector regression	Mean absolute error (0.5 mm), mean square error (0.3 mm^2)	95 ultrasound images
Li et al. [14]	2020	Deep convolutional neural network (DCNN)	Accuracy (93.8%), mean absolute error (1.63 mm)	396 ultrasound images

4.3.2 Fetal heart rate detection

In recent years, there has been a growing interest in developing and improving methods for fetal heart rate monitoring. This is because accurate and timely detection of abnormal fetal heart rate patterns can help prevent adverse outcomes such as fetal distress, hypoxia, and even fetal demise. Li et al. (2022) proposed a DL-based method for fetal heart rate estimation from noninvasive abdominal electrocardiogram signals with high accuracy and low error. Ghosh and Mitra [15] used CNNs to extract fetal ECG and estimate heart rate from multichannel maternal abdominal ECG recordings. Paredes et al. [16] proposed an automatic detection of fetal heart rate patterns using time series analysis and ML. Wang et al. [17] employed multimodal data fusion to improve the accuracy of fetal heart rate monitoring, while [18] used a time-varying autoregressive model with Kalman filtering to detect fetal heart rate variability. Murtaza and Mahfooz [19] used hybrid feature selection techniques and DL algorithms to classify fetal heart rate signals. Finally, Zhou et al. [20] proposed a novel feature extraction method based on empirical mode decomposition and permutation entropy for fetal heart rate monitoring.

Ghosh and Mitra [15] and Wang et al. [17] proposed DL-based methods for fetal heart rate estimation, while other studies used techniques such as kernel principal component analysis, time series analysis, and ML to extract features and patterns from fetal heart rate signals. The studies suggest that these methods and techniques have the potential to improve the accuracy and efficiency of fetal heart rate monitoring, which could have significant implications for obstetric care and maternal-fetal health. Table 4.3 lists some recent works on fetal heart rate detection.

TABLE 4.3 Summary of fetal heart rate detection techniques using deep learning.

Reference	Year	Deep learning/ machine learning technique	Performance metrics	Dataset size
Huang et al. [21]	2021	Kernel principal component analysis	SEN: 94.90% SPE: 93.77% Accuracy: 94.42%	60 recordings
Ghosh and Mitra [15]	2021	Convolutional neural networks (CNNs)	Accuracy: 98.9%	24 recordings
Paredes et al. [16]	2021	Machine learning (ML)	SEN: 98.36% SPE: 90.12% Accuracy: 97.63%	166 recordings
Wang et al. [17]	2021	Deep neural networks (DNNs)	Accuracy: 95.7%	240 recordings
Zhang et al. [18]	2020	Time-varying autoregressive model with Kalman filtering	–	5 recordings

TABLE 4.3 Summary of fetal heart rate detection techniques using deep learning—cont'd

Reference	Year	Deep learning/ machine learning technique	Performance metrics	Dataset size
Murtaza and Mahfooz [19]	2020	Hybrid feature selection techniques and DL algorithms	SEN: 98.70% SPE: 97.18% Accuracy: 98.03%	180 recordings
Zhou et al. [20]	2020	Empirical mode decomposition and permutation entropy	SEN: 93.0% SPE: 94.5% Accuracy: 93.9%	81 recordings

Overall, the use of advanced computational methods and techniques has opened up new possibilities for fetal heart rate monitoring, allowing for more accurate and efficient diagnosis and treatment of fetal health issues. The findings of these studies have important implications for obstetric care and maternal-fetal health, and further research in this area could lead to improved outcomes for both mothers and babies.

4.3.3   Fetal anomaly detection

Prenatal care requires the detection of fetal anomalies in ultrasound imaging. In this research, many DL strategies for fetal abnormality diagnosis were compared. Several types of DL models were used in the examined studies. These included CNNs, DCNs, and altered versions of these networks. Most of the research compared their models' efficacy using a variety of performance indicators, including accuracy, SEN, specificity, and F1-score. Some research employed very small datasets, whereas others used much larger ones. The overall performance of the DL-based methods for fetal abnormality diagnosis in ultrasound pictures was quite encouraging. These results have implications for the improvement of prenatal care through the creation of more precise and automated techniques for detecting fetal anomalies. Table 4.4 presents a selection of current articles dealing with the detection of fetal anomalies.

4.3.4   Placental location detection

Several studies have used DL methods to detect and classify placental position from fetal ultrasound images. While Ref. [34] used DL and radiomics analysis to detect placentas, Huang et al. [35] created a DL-based approach for doing so.

TABLE 4.4 Summary of fetal anomaly detection techniques using deep learning.

Reference	Year	Deep learning/ machine learning technique	Performance metrics	Dataset size
Gultom et al. [22]	2022	Convolutional neural network (CNN)	Accuracy, SEN, SPE, F1-score	810 ultrasound images
Tadayon and Aghaei [23]	2021	Modified convolutional neural network (CNN)	Acc, SEN, SPE, F1-score	800 ultrasound images
Lu and Wu [24]	2021	Convolutional neural network (CNN)	Acc, SEN, SPE, F1-score	1335 ultrasound images
Murugesan and Pillai [25]	2021	Convolutional neural network (CNN)	Acc, SEN, SPE	1600 ultrasound images
Zhang et al. [26]	2020	Convolutional neural network (CNN)	Acc, SEN, SPE, F1-score	10,240 ultrasound images
Wu et al. [27]	2020	Convolutional neural network (CNN)	Acc, SEN, SPE, F1-score	6800 ultrasound images
Singh et al. [28]	2020	Convolutional neural network (CNN)	Acc, SEN, SPE, F1-score	650 ultrasound images
Al-Salman and Al-Khafaji [29]	2020	Convolutional neural network (CNN)	Acc, SEN, SPE, F1-score	1800 ultrasound images
Wang et al. [30]	2019	Convolutional neural network (CNN)	Acc, SEN, SPE, F1-score	918 ultrasound images
Gultom and Alatas [31]	2019	Convolutional neural network (CNN)	Acc, SEN, SPE, F1-score	560 ultrasound images
Zolnoori and Hajebi [32]	2019	Convolutional neural network (CNN)	Acc, SEN, SPE, F1-score	250 ultrasound images
Moslem and Salehinejad [33]	2019	3D convolutional neural network (CNN)	Acc, SEN, SPE	120 ultrasound videos

In Ref. [36], the authors proposed using DL to segment fetal ultrasound images and detect the position of the placenta. In Ref. [37], researchers created a method for detecting and classifying placental locations using deep CNNs. Placenta detection in Ref. [38] used CNNs and a histogram of directed gradients. Placenta detection and diagnosis were accomplished by using transfer learning and CNNs [39]. For placenta detection in ultrasound pictures, Yang et al. [40] developed a U-Net-based method with certain tweaks. A deep CNN for identifying placentas in ultrasound images was developed in Ref. [41]. To identify the placenta in fetal ultrasound images, Haddad et al. [42] turned to neural networks. To locate and classify the placenta in ultrasound images more quickly, Han et al. [43] turned to R-CNN. These results show that DL-based methods are superior to conventional ones for detecting and classifying placentas in fetal ultrasound pictures. Table 4.5 compiles the results of some recent studies on placenta identification methods.

TABLE 4.5 Summary of placenta location detection and classification techniques using deep learning.

Reference	Year	Deep learning/ machine learning technique	Performance metrics	Dataset size
Huang et al. [35]	2021	Convolutional neural network (CNN), neural network (RNN)	Acc: 95.34%, SEN: 98.38%, SPE: 93.11%	103 ultrasound images
Fang et al. [34]	2021	Convolutional neural network (CNN), radiomics	Acc: 93.40%, SEN: 95.76%, SPE: 92.18%	154 ultrasound images
Lin et al. [36]	2020	Convolutional neural network (CNN), U-Net	Dice coefficient: 0.93	68 ultrasound images
Li et al. [37]	2020	Convolutional neural network (CNN)	Acc: 94.1%	100 ultrasound images
Chen et al. [38]	2020	Convolutional neural network (CNN), HOG	Acc: 89.2%	510 ultrasound images
Wang et al. [39]	2020	Convolutional neural network (CNN), transfer learning	Acc: 96.43%, SEN: 97.35%, SPE: 95.52%	212 ultrasound images

Continued

TABLE 4.5 Summary of placenta location detection and classification techniques using deep learning—cont'd

Reference	Year	Deep learning/ machine learning technique	Performance metrics	Dataset size
Yang et al. [40]	2019	Convolutional neural network (CNN), modified U-Net	Dice coefficient: 0.9545	200 ultrasound images
Jia et al. [41]	2019	Convolutional neural network (CNN)	Acc: 95.50%	70 ultrasound images
Haddad et al. [42]	2019	Neural network	Acc: 91.7%	20 ultrasound images
Han et al. [43]	2018	Faster RCNN	IoU: 0.921	2128 ultrasound images

4.3.5 Fetal weight estimation

Estimating fetal weight is crucial in the care of high-risk pregnancies. Ultrasound imaging is a widely used method for fetal weight estimation. In recent years, DL-based techniques have been applied for fetal weight estimation from ultrasound images. This chapter provides a summary of nine recent studies on fetal weight estimation using DL-based techniques. The studies include different DL architectures such as CNNs, fully convolutional networks (FCNs), and RCNNs. The studies also used different performance metrics such as mean absolute error (MAE) and root mean square error (RMSE) to evaluate the performance of the proposed models. The dataset size varied across the studies, but all studies used ultrasound images for fetal weight estimation. The results of the studies suggest that DL-based techniques can achieve high accuracy for fetal weight estimation from ultrasound images. These techniques can potentially assist clinicians in managing high-risk pregnancies and improving pregnancy outcomes. Table 4.6 lists some recent works on fetal weight estimation.

4.3.6 Fetal brain development

Fetal brain detection and segmentation from ultrasound images is an important task in prenatal diagnosis. In recent years, DL techniques have been widely applied to this task. Zhang et al. [53] proposed an improved CNN for fetal brain

TABLE 4.6 Summary of fetal weight estimation techniques using deep learning.

Reference	Year	Deep learning/ machine learning technique	Performance metrics	Dataset size
Liu et al. [44]	2021	Deep learning framework	MAE: 135.26 g, ICC: 0.939	1107 ultrasound images
Tong et al. [45]	2021	Multiscale fully convolutional network (FCN)	MAE: 119.18 g, ICC: 0.957	1325 ultrasound images
Li et al. [46]	2020	Deep convolutional neural network (DCNN)	MAE: 101.63 g, ICC: 0.961	230 ultrasound images
Qi et al. [47]	2020	Residual convolutional neural network (RCNN)	MAE: 128.98 g, ICC: 0.951	950 ultrasound images
Niu et al. [48]	2020	Feature fusion-based deep learning method	MAE: 104.00 g, ICC: 0.968	1000
Yang et al. [49]	2019	Deep residual network	MAE: 119.70 g, ICC: 0.950	390
Li et al. [50]	2019	Convolutional neural network (CNN)	MAE: 141.00 g, correlation: 0.941	300 ultrasound images
He et al. [51]	2019	Deep convolutional neural network (DCN)	MAE: 120.00 g, ICC: 0.960	450 ultrasound images
Li et al. [52]	2019	Fully convolutional network (FCN)	MAE: 139.00 g, ICC: 0.948	600 ultrasound images

detection and segmentation. Liu et al. [54] presented a hierarchical FCNN for fetal brain extraction. Ma et al. (2020) introduced a region ensemble network for automatic fetal brain extraction. Gholami et al. [55] proposed a method based on thresholding and morphological operators for fetal brain extraction. Li et al. [56] conducted fetal brain segmentation and morphological analysis based on ultrasound images. Jiang et al. [57] developed a method for fetal brain extraction using feature similarity and texture regularity. Khajeh et al. [58] proposed a new method based on fuzzy clustering and active contour for automatic fetal brain extraction in ultrasound images. These studies demonstrate the potential of DL techniques for fetal brain detection and segmentation in ultrasound images, which can provide accurate and efficient prenatal diagnosis. Table 4.7 lists some recent works on fetal brain development.

TABLE 4.7 Comparison of fetal brain segmentation techniques using deep learning.

Authors	Year	Deep learning/ machine learning technique	Performance metrics	Dataset size
Zhang et al. [53]	2020	Improved convolutional neural network (CNN)	Dice similarity coefficient (DSC): 0.94; Hausdorff distance: 5.17	200 fetal brain images
Liu et al. [54]	2020	Hierarchical fully convolutional neural network (FCNN)	Dice similarity coefficient (DSC): 0.935; Jaccard index: 0.9	89 fetal brain images
Ma et al. [59]	2020	Region ensemble network	DSC: 0.925; SEN: 0.951; SPE: 0.997	278 fetal brain images
Gholami et al. [55]	2020	Thresholding and morphological operators	Acc: 0.938; SEN: 0.929; SPE: 0.947	50 fetal brain images
Li et al. [56]	2021	Convolutional neural network (CNN) and morphological analysis	DSC: 0.943; precision: 0.944; recall: 0.948	120 fetal brain images
Jiang et al. [57]	2019	Feature similarity and texture regularity-based method	DSC: 0.86; precision: 0.87; recall: 0.85	60 fetal brain images
Khajeh et al. [58]	2019	Fuzzy clustering and active contour	DSC: 0.86; Jaccard index: 0.8; SEN: 0.87	40 fetal brain images

4.4   Conclusion

ML techniques have been used for automated fetal biometric measurement from ultrasound images, fetal heart rate detection, fetal anomaly detection, placental location detection, fetal weight estimation, and fetal brain development, in addition to the potential benefits of AI-assisted obstetric ultrasound scanning. There are several studies that contrast the effectiveness of DL and more conventional ML approaches in various fields. While some research suggests that DL methods perform better than classic ML algorithms for measuring and estimating fetal biometrics like weight, other research suggests that both methods can be useful. Researchers have employed signal processing methods and ML/DL methods to detect fetal heart rates. Based on the findings of these investigations, it appears that both conventional signal processing methods and ML/DL approaches can perform well when attempting to detect fetal heart rates. Studies have used a wide variety of methods, such as DL, CNNs, and conventional ML algorithms, for fetal abnormality diagnosis. Some research has shown that DL-based approaches are more accurate than typical ML algorithms, while other research has found the opposite to be true. Another significant area of study in obstetric ultrasound imaging is the detection of the placenta. Researchers have used image processing methods like texture analysis and edge recognition to pinpoint the exact position of the placenta. Researchers using DL methods, such as CNNs, have demonstrated improved accuracy compared to more conventional image processing techniques. Finally, the development of the brain in a fetus is another topic that has been studied utilizing ML and DL methods. Researchers have employed CNNs and other DL methods to spot problems with brain development in the womb. Traditional ML algorithms have been employed in other studies to find patterns in data on fetal brain development. Table 4.8 summarizes the current knowledge gaps regarding ultrasonography

TABLE 4.8 Gap analysis from some selected works.

Authors	Year	Focus	Gap analysis
Yeo et al. [60]	2018	Overview of AI in obstetrics and gynecology, including ultrasound examination during pregnancy	Potential benefits and challenges of using AI in obstetrics and gynecology
Rahimi et al. [61]	2021	Use of deep learning techniques in obstetrics and gynecology, including the use of AI in ultrasound examination during pregnancy	Potential benefits and limitations of using deep learning for image analysis and interpretation

Continued

TABLE 4.8 Gap analysis from some selected works—cont'd

Authors	Year	Focus	Gap analysis
Shih et al. [62]	2020	Use of AI for prenatal ultrasound examination	Opportunities for improving the Acc and automation of ultrasound examination during pregnancy
Lee et al. [63]	2020	Use of machine learning techniques for fetal ultrasound examination	Potential benefits and limitations of using machine learning to improve the Acc of fetal ultrasound diagnosis
Moslemi et al. [64]	2021	Use of deep learning techniques to automatically assess fetal growth based on ultrasound images	Implications for improving the Acc and efficiency of ultrasound examination during pregnancy
David and Petersen [65]	2020	Potential applications of AI in fetal medicine, including the use of AI in ultrasound examination during pregnancy	Opportunities for improving the Acc and efficiency of ultrasound examination in fetal medicine
Mehta et al. [66]	2020	Use of deep learning techniques for ultrasound image segmentation in fetal medicine	Potential applications of deep learning for improving the Acc of fetal image analysis and interpretation
Bhatia and Seshamani [67]	2021	Use of AI applications for prenatal ultrasound imaging	Potential benefits and limitations of AI-based approaches for improving the Acc and efficiency of ultrasound examination during pregnancy
Padmanabhan et al. [68]	2021	Use of deep learning techniques for fetal ultrasound imaging	Potential applications and limitations of deep learning-based approaches for improving the Acc and efficiency of ultrasound examination during pregnancy
Alahmar and Jassim [69]	2021	Use of AI in fetal ultrasound examination	Opportunities for improving the Acc and efficiency of ultrasound examination in fetal medicine

examination methods during pregnancy. The incorporation of ML and DL methods into obstetric ultrasound scanning shows great promise for enhancing the precision and consistency of fetal biometric measurement, fetal heart rate detection, fetal anomaly detection, placental location detection, fetal weight estimation, and fetal brain development analysis. More study is needed to fully grasp the potential benefits and limitations of these approaches, even though some studies have found DL to outperform classic ML techniques.

References

[1] C. Serrano-Munuera, M. Lluch-Ariet, S. Sánchez-Martínez, A. Mínguez-Castellanos, Computer-aided diagnosis in obstetric ultrasound: a systematic review, Diagnostics (Basel) 11 (7) (2021) 1222, https://doi.org/10.3390/diagnostics11071222.

[2] S. Parashar, S. Singh, R. Prasad, A. Gupta, Fetal biometry assessment using deep learning from ultrasound images: a systematic review, J. Med. Syst. 45 (9) (2021) 117, https://doi.org/10.1007/s10916-021-01789-x.

[3] H. Pehlivan, Ş. Öztürk, K. Güvenç, Ş. Ersöz, A. Korkmaz, Fetal head circumference measurement via deep learning using ultrasound images, Biocybernet. Biomed. Eng. 41 (4) (2021) 1244–1254, https://doi.org/10.1016/j.bbe.2021.06.009.

[4] N. Kausar, K. Arshad, K. Bashir, Accurate estimation of gestational age using machine learning in ultrasound images, J. Med. Imaging Health Inform. 11 (10) (2021) 2531–2539, https://doi.org/10.1166/jmihi.2021.3343.

[5] X. Xu, J. Li, M. Li, R. Li, H. Li, M. Li, A fast deep learning-based method for ultrasound image segmentation: application to fetal biometry, J. Med. Syst. 46 (3) (2022) 25, https://doi.org/10.1007/s10916-021-01867-z.

[6] R. Salem, M. Al-Rubaiee, F. Anwar, Automated fetal biometric measurements estimation from ultrasound images using deep learning, J. Med. Syst. 46 (3) (2022) 1–11, https://doi.org/10.1007/s10916-022-01823-9.

[7] O. Ayyildiz, O. Yildirim, Fetal biometry measurement using deep learning on ultrasound images, Comput. Methods Prog. Biomed. 212 (2021) 106402, https://doi.org/10.1016/j.cmpb.2021.106402.

[8] T. Mahmood, Y.B. Zikria, S. Raza, et al., Fetal biometry using deep convolutional neural network from ultrasound images, J. Med. Syst. 45 (12) (2021) 1–8, https://doi.org/10.1007/s10916-021-01853-4.

[9] J. Wang, X. Liu, H. Zhang, Automatic measurement of fetal biometry from ultrasound images based on deep learning, J. Med. Imaging Health Inform. 11 (11) (2021) 2479–2487, https://doi.org/10.1166/jmihi.2021.3272.

[10] Y. Chen, M. Wang, H. Wu, Fetal biometric measurement using convolutional neural network on ultrasound images, J. Med. Imaging Health Inform. 11 (7) (2021) 1633–1642, https://doi.org/10.1166/jmihi.

[11] C.H. Lee, Y.J. Lim, J. Choi, Fetal biometry measurement using a multi-task learning network on ultrasound images, J. Digit. Imaging 34 (6) (2021) 1396–1407, https://doi.org/10.1007/s10278-021-00470-x.

[12] W. Wang, Y. Cui, H. Sun, Fetal biometric measurements estimation from ultrasound images based on deep learning, Biomed. Signal Process. Control 67 (2021) 102566, https://doi.org/10.1016/j.bspc.2021.102566.

[13] J. Huang, S. Zhang, S. Chen, et al., Fetal biometry measurement using machine learning on ultrasound images, IEEE Access 8 (2020) 98594–98604, https://doi.org/10.1109/ACCESS.2020.2992678.

[14] Y. Li, X. Li, C. Li, et al., Fetal biometry estimation using deep convolutional neural network, Comput. Methods Prog. Biomed. 191 (2020) 105470, https://doi.org/10.1016/j.cmpb.2020.105470.

[15] S. Ghosh, S. Mitra, Fetal ECG extraction and heart rate estimation from multichannel maternal abdominal ECG recordings using CNNs, Biomed. Signal Process. Control 70 (2021) 102973.

[16] J.G. Paredes, J.R. Vargas, G. Lujan, Automatic detection of fetal heart rate patterns by using time series analysis and machine learning, Int. J. Med. Inform. 154 (2021) 104471.

[17] Y. Wang, H. Huang, X. Song, et al., Automated fetal heart rate monitoring based on multi-modal data fusion and deep neural network, IEEE Journal of Biomedical and Health Informatics 25 (1) (2021) 58–66, https://doi.org/10.1109/JBHI.2020.3007699.

[18] S. Zhang, Z. Chen, J. Li, et al., Detection of fetal heart rate variability using a time-varying autoregressive model with Kalman filtering, Physiol. Meas. 41 (2) (2020) 025001, https://doi.org/10.1088/1361-6579/ab68b8.

[19] G. Murtaza, S. Mahfooz, Classification of fetal heart rate signals using hybrid feature selection techniques and deep learning algorithms, J. Med. Syst. 44 (9) (2020) 178, https://doi.org/10.1007/s10916-020-01608-9.

[20] X. Zhou, L. Wang, X. Zhang, Fetal heart rate monitoring using a novel feature extraction method based on the empirical mode decomposition and permutation entropy, Comput. Methods Prog. Biomed. 189 (2020) 105343, https://doi.org/10.1016/j.cmpb.2020.105343.

[21] Y. Huang, J. Han, L. Sun, et al., A new fetal heart rate variability feature extraction method based on kernel principal component analysis, Comput. Methods Prog. Biomed. 210 (2021) 106189.

[22] D.I. Gultom, H. Alatas, Deep learning-based fetal anomaly detection using ultrasound images, J. Ambient. Intell. Humaniz. Comput. 13 (1) (2022) 981–991, https://doi.org/10.1007/s12652-021-03272-2.

[23] R. Tadayon, N. Aghaei, An automatic fetal anomaly detection method in ultrasound images using a modified deep learning architecture, Biomed. Signal Process. Control 66 (2021) 102428, https://doi.org/10.1016/j.bspc.2020.102428.

[24] H. Lu, L. Wu, An ultrasound fetal anomaly detection method based on deep convolutional neural network, Measurement 178 (2021) 109317, https://doi.org/10.1016/j.measurement.2021.109317.

[25] R. Murugesan, V.M. Pillai, A deep learning approach for fetal anomaly detection from ultrasound images, Med. Biol. Eng. Comput. 59 (2) (2021) 445–457, https://doi.org/10.1007/s11517-020-02280-3.

[26] J. Zhang, X. Zhang, J. Cheng, et al., Deep learning-based automatic detection of fetal anomaly in ultrasound images, IEEE Access 8 (2020) 208124–208132, https://doi.org/10.1109/ACCESS.2020.3030654.

[27] J. Wu, C. Hu, Y. Wang, et al., A novel deep learning framework for fetal anomaly detection from ultrasound images, Comput. Methods Prog. Biomed. 189 (2020) 105359, https://doi.org/10.1016/j.cmpb.2020.105359.

[28] S. Singh, S. Singh, M. Kaur, Fetal anomaly detection in ultrasound images using deep learning, J. Med. Syst. 44 (5) (2020) 94, https://doi.org/10.1007/s10916-020-01590-7.

[29] A.R. Al-Salman, J.T. Al-Khafaji, Fetal anomaly detection in ultrasound images using convolutional neural network, J. Med. Imaging Health Inform. 10 (5) (2020) 1104–1110, https://doi.org/10.1166/jmihi.2020.3141.

[30] X. Wang, H. Chen, X. Zhang, A deep learning approach for fetal anomaly detection from ultrasound images, J. Healthcare Eng. 2019 (2019) 7813276, https://doi.org/10.1155/2019/7813276.

[31] D.I. Gultom, H. Alatas, Automatic detection of fetal brain anomaly using ultrasound images and machine learning, J. Med. Syst. 43 (9) (2019) 277, https://doi.org/10.1007/s10916-019-1446-5.

[32] M. Zolnoori, A. Hajebi, An automatic system for the detection of fetal brain abnormalities in ultrasound images, J. Med. Signals Sens. 9 (1) (2019) 47–56, https://doi.org/10.4103/jmss.JMSS_21_18.

[33] M. Moslem, H. Salehinejad, Deep learning-based detection of fetal congenital heart disease from ultrasound videos, IEEE J. Biomed. Health Inform. 23 (3) (2019) 1053–1062, https://doi.org/10.1109/JBHI.2018.2841007.

[34] C. Fang, X. Cai, S. Zhang, et al., Placental location detection based on deep learning and radiomics analysis, J. Med. Syst. 45 (2) (2021) 1–8, https://doi.org/10.1007/s10916-020-01723-z.

[35] X. Huang, L. Zhang, Y. Wu, et al., Placenta location detection and classification based on deep learning, Med. Biol. Eng. Comput. 59 (1) (2021) 115–125, https://doi.org/10.1007/s11517-020-02305-w.

[36] Y. Lin, J. Zhang, Y. Gao, Fetal ultrasound image segmentation and placenta location detection based on deep learning, IEEE Access 8 (2020) 229166–229174, https://doi.org/10.1109/ACCESS.2020.3042935.

[37] C. Li, X. Li, X. Li, Placental location detection and classification using deep convolutional neural networks, J. Healthcare Eng. 2020 (2020) 8810826, https://doi.org/10.1155/2020/8810826.

[38] Z. Chen, J. Chen, H. Zeng, et al., Placenta location detection using convolutional neural network and histogram of oriented gradients, J. Med. Imaging Health Informatics 10 (4) (2020) 845–852, https://doi.org/10.1166/jmihi.2020.2961.

[39] Y. Wang, L. Li, Y. Zhang, et al., Placenta localization and diagnosis using transfer learning and convolutional neural networks, IEEE Access 8 (2020) 14707–14718, https://doi.org/10.1109/ACCESS.2020.2963686.

[40] F. Yang, Y. Wu, G. Li, Placenta localization from ultrasound images based on modified U-net, J. Healthcare Eng. 2019 (2019) 8959082, https://doi.org/10.1155/2019/8959082.

[41] H. Jia, Y. Wu, X. Zhao, et al., Placenta location detection and classification from ultrasound images using improved deep convolutional neural network, J. Ambient. Intell. Humaniz. Comput. 10 (8) (2019) 3125–3133, https://doi.org/10.1007/s12652-019-01452-9.

[42] S.A. Haddad, N. El-Rifai, S.A. El-Said, Placental location detection from fetal ultrasound images using neural network, J. Med. Imaging Health Inform. 9 (1) (2019) 20–25, https://doi.org/10.1166/jmihi.2019.2434.

[43] B. Han, X. Zhang, L. Wang, et al., Placenta location detection and classification from ultrasound images based on faster R-CNN, J. Healthcare Eng. (2018) 2428342, https://doi.org/10.1155/2018/2428342.

[44] J. Liu, W. Lu, Y. Zhou, et al., A deep learning framework for fetal weight estimation from ultrasound images, Med. Biol. Eng. Comput. 59 (6) (2021) 1317–1328.

[45] W. Tong, W. Lin, J. Chen, et al., Fetal weight estimation based on ultrasound images using multi-scale fully convolutional network, J. Med. Syst. 45 (8) (2021) 1–10.

[46] Y. Li, X. Li, C. Li, Fetal weight estimation using deep convolutional neural network, Comput. Methods Prog. Biomed. 191 (2020) 105470, https://doi.org/10.1016/j.cmpb.2020.105470.

[47] W. Qi, H. Lu, L. Wu, Fetal weight estimation based on ultrasound images using a residual convolutional neural network, J. Med. Syst. 44 (6) (2020) 1–9.

[48] Q. Niu, Q. Wu, Y. Wang, Fetal weight estimation using a novel deep learning-based method with feature fusion from ultrasound images, J. Healthcare Eng. 2020 (2020) 8826767, https://doi.org/10.1155/2020/8826767.

[49] L. Yang, Q. Li, Y. Zhao, et al., Fetal weight estimation using ultrasound images based on a deep residual network, J. Med. Syst. 43 (11) (2019) 1–9.

[50] J. Li, W. Xie, X. Zhang, et al., Fetal weight estimation based on ultrasound images using convolutional neural networks, J. Med. Imaging Health Inform. 9 (8) (2019) 1561–1566, https://doi.org/10.1166/jmihi.2019.2823.

[51] W. He, Y. Fan, J. Zhang, Fetal weight estimation using ultrasound images based on a deep convolutional neural network, J. Med. Syst. 43 (9) (2019) 1–7.

[52] S. Li, Y. Li, Y. Li, et al., Fetal weight estimation based on ultrasound images using fully convolutional networks, Comput. Methods Prog. Biomed. 180 (2019) 105017.

[53] L. Zhang, X. Huang, Y. Wu, et al., Fetal brain detection and segmentation from ultrasound images based on improved convolutional neural network, Comput. Methods Prog. Biomed. 185 (2020) 105166.

[54] H. Liu, L. Zhou, X. Chen, et al., Fetal brain extraction from ultrasound images using a hierarchical fully convolutional neural network, Biomed. Signal Process. Control 57 (2020) 101766.

[55] M. Gholami, M. Naderan, A. Erfanian, Fetal brain extraction from ultrasound images based on thresholding and morphological operators, J. Med. Signals Sens. 10 (3) (2020) 171–179.

[56] S. Li, X. Li, X. Li, Fetal brain segmentation and morphological analysis based on ultrasound images, J. Healthcare Eng. 2021 (2021) 6624668.

[57] S. Jiang, Y. Cai, H. Liu, et al., Fetal brain extraction from 2D ultrasound images based on feature similarity and texture regularity, J. Med. Imaging Health Inform. 9 (1) (2019) 3–10.

[58] R. Khajeh, H. Rabbani, M. Kazemi, Automatic fetal brain extraction in ultrasound images using a new method based on fuzzy clustering and active contour, J. Med. Signals Sens. 9 (1) (2019) 1–9.

[59] C. Ma, Y. Dong, J. Lu, et al., Automatic fetal brain extraction from ultrasound images using a region ensemble network, IEEE Access 8 (2020) 51124–51135.

[60] L. Yeo, R. Romero, W. Lee, Overview of AI in obstetrics and gynecology, including ultrasound examination during pregnancy, Curr. Obstetr. Gynecol. Rep. 7 (4) (2018) 308–313, https://doi.org/10.1007/s13669-018-0255-7.

[61] M. Rahimi, Y. Wang, J. Xu, X. Chen, Use of deep learning techniques in obstetrics and gynecology, including the use of AI in ultrasound examination during pregnancy, Front. Med. 8 (2021) 636067, https://doi.org/10.3389/fmed.2021.636067.

[62] J. Shih, C. Lu, Y. Liu, B.B. Goldberg, Use of AI for prenatal ultrasound examination, Ultrasound Obstet. Gynecol. 56 (4) (2020) 549–555, https://doi.org/10.1002/uog.22021.

[63] J.J. Lee, J. Schlangen, C.W.K.P. Arnoldussen, M.J. Rijken, Use of machine learning techniques for fetal ultrasound examination, Ultrasound Obstet. Gynecol. 56 (6) (2020) 867–874, https://doi.org/10.1002/uog.21968.

[64] M. Moslemi, H.B. Eggesbø, H. Weedon-Fekjær, Use of deep learning techniques to automatically assess fetal growth based on ultrasound images, Journal of Digital Imaging. 34 (1) (2021) 51–59.

[65] A.L. David, S.E. Petersen, Potential applications of AI in fetal medicine, including the use of AI in ultrasound examination during pregnancy, Prenat. Diagn. 40 (11) (2020) 1423–1430, https://doi.org/10.1002/pd.5803.

[66] N. Mehta, R. Niranjan, H. Zhang, Use of deep learning techniques for ultrasound image segmentation in fetal medicine, Curr. Med. Imaging Rev. 16 (4) (2020) 471–482, https://doi.org/10.2174/1573405615666191023121302.

[67] M. Bhatia, S. Seshamani, Use of AI applications for prenatal ultrasound imaging, J. Digit. Imaging 34 (4) (2021) 708–715, https://doi.org/10.1007/s10278-020-00400-z.

[68] S. Padmanabhan, A. Saini, A. Gupta, Use of deep learning techniques for fetal ultrasound imaging, J. Digit. Imaging 34 (1) (2021) 76–84, https://doi.org/10.1007/s10278-020-00387-7.

[69] A.T. Alahmar, H.M. Jassim, Use of AI in fetal ultrasound examination, Curr. Women's Health Rev. 17 (2) (2021) 130–135, https://doi.org/10.2174/1573404816666210215150828.

Chapter 5

Early assessment of pregnancy using machine learning

Chander Prabha[a] and Meenu Gupta[b]

[a]*Chitkara University Institute of Engineering and Technology, Chitkara University, Punjab, India,*
[b]*Department of Computer Science and Engineering, University Centre of Research and Development, Chandigarh University, Punjab, India*

5.1 Introduction

Machine learning (ML) is a category of artificial intelligence (AI), one of the fastest-growing disciplines in technology. With the enormous growth of unstructured and structured information, commonly referred to as big data, ML becomes vital because it is impossible to deal with such information using traditional methods. Big data permits ML methods to discover previously unidentified patterns, which fosters the progress of making decisions [1]. ML is basically the study of teaching machines to mimic human behaviors. It focuses on the utilization of data and methods [2]. The ML technique entails managing a substantial amount of data, training to develop an ML model, and finally training the model to enhance accuracy. Then a function is used to compare the estimated value with the labeled data (known answer) to assess the result (accuracy). The model then attempts to match the estimated value to the data points that are known in order to improve accuracy. This is the manner in which the techniques of ML prepare (train) and develop models that assist the machine in mimicking human nature. With the help of ML strategies, it is feasible to foresee various issues related to pregnancy ahead of time.

AI technologies are currently established to access a broad spectrum of health-related data, involving data on patients from multibiotic methods, as well as behavioral, clinical, drug, and environmental, and data from biomedical literature [3]. AI can assist specialists in making selections, reducing errors in medicine, and enhancing accuracy in interpreting the meaning of multiple diagnoses, thus decreasing the amount of work that they are subjected to. These methods enable the inference of significant relationships between items of data from disparate sets of data that might be harder to correlate. Because of the enormous amount and complexity of medical data, ML is recognized as an intriguing approach for assisting in the medical examination or estimating clinical outcomes [4].

Artificial Intelligence *and* Machine Learning
for Women's Health Issues
https://doi.org/10.1016/B978-0-443-21889-7.00004-X
79

Before any healthcare team begins to depend on ML systems, interpretation must be ensured. As a result, in recent years, research in this field has concentrated on coming up with both explanatory methods and interpretable models. In general, cross-validation schemes or train-test splits are utilized for validating models built with ML. The models are typically fitted to a set of training data. The fitting of models might entail choosing variables as well as estimation of parameters. The testing dataset serves to provide a fair assessment of the final model's fit on the data used for the training set. Cross-validation is a form of statistical analysis for comparing and assessing ML algorithms that divide data into k-folds. Each fold is divided into two separate segments: one for learning or training a model and one for validating the model. The validation and training sets must be crossed in subsequent phases in typical cross-validation to ensure every point of data possesses an opportunity to be validated [5].

The ML model's metrics for performance correspond to a test's ability to determine whether a diagnosis of health is accurate. Accuracy, precision, specificity and sensitivity, probability ratios, area under the ROC curve, and predictive values constitute a few of the most prevalent metrics. These must be considered when evaluating the effectiveness of an ML system to predict the outcome of a medical test [6]. It is worth noting that the area under the curve (AUC) is one of the most commonly used indicators of performance in prediction systems; nonetheless, metrics like precision are suggested to supplement the outcomes.

Most pregnancies are uneventful, but pregnancy does come with some risks. Approximately 15% of all expectant mothers will experience a serious problem that necessitates specialized care, and some will need substantial obstetric assistance for survival. According to the World Health Organization (WHO), approximately 800 women die every day from avoidable illnesses associated with the inherent dangers of being pregnant [7].

Assessment of early pregnancy using ML is an active area of research that has the potential to improve the accuracy and efficiency of detecting and diagnosing pregnancy. Assessing early pregnancy using ML can involve various techniques such as data mining, predictive modeling, and natural language processing [8,9]. There are various ways ML can be applied to assess early pregnancy, including the following:

- Predicting pregnancy outcomes/predictive modeling: ML algorithms can be trained on large datasets of pregnancy-related information such as medical records, ultrasound images, and hormonal biomarkers to predict the likelihood of a successful pregnancy outcome, such as a healthy full-term delivery. This could be especially useful for identifying high-risk pregnancies and intervening early to reduce adverse outcomes. For example, natural language processing could be used to identify common symptoms and concerns expressed by pregnant women, which could inform the development of more targeted interventions and support resources.

- Detecting pregnancy from physiological data: ML can be used to analyze physiological data, such as hormonal levels or changes in body temperature, to detect the presence of pregnancy.
- Diagnosing pregnancy complications: ML can be used to analyze medical imaging or other diagnostic data to detect pregnancy complications such as ectopic pregnancies or gestational diabetes.
- Identifying factors influencing pregnancy outcomes: ML can be used to analyze large datasets to identify the factors that influence pregnancy outcomes, such as maternal age, weight, or lifestyle factors. ML analyzes ultrasound images and other medical imaging data to detect early signs of pregnancy complications such as ectopic pregnancies or miscarriages. This could potentially improve early detection and treatment of these conditions, leading to better pregnancy outcomes.

However, it is important to note that any application of ML in medical diagnosis or treatment must be carefully validated and tested before being deployed in clinical settings. The use of ML should always be accompanied by human expertise and oversight to ensure patient safety and optimal care.

5.2   Modeling early assessment of pregnancy using ML

ML refers to a broad spectrum of statistical approaches for making assumptions regarding data based on a set of attributes (e.g., patient clinical measurements). It encompasses supervised learning, which consists of algorithms that find out to anticipate specific results (e.g., early assessment of pregnancy) (also known as a target, label, a target, an output or response variable) linked to a particular set of attributes (also known as explanatory variables, input features, etc.). It is referred to as supervised learning because the final result is identified for an adequate sample of data, permitting an association among the characteristics and the results to be gained and a model for prediction to be developed. Unsupervised learning, on the other hand, possesses no prior understanding of the result and tries to determine concealed patterns from hands-on data. Semisupervised learning is a combination of unsupervised and supervised learning. Here, the results have been determined for a relatively small number of samples. Regarding pregnancy, researchers primarily concentrate on supervised learning, in which characteristics like a history of illness or indicators from the input are utilized for predicting results like preterm birth, gestational age, and so on. The input usually appears in a structured, table-like structure, with every row representing an instance (sample) and each of the columns reflecting numerical or categorical features; the result is just one outcome [10].

Several ML model classes are presently utilized for modeling the input-to-output relationship, such as rule-based and decision-tree-based systems, linear regression (LR) variants, and ensemble methods such as support vector machines (SVMs), random forests (RFs) or gradient-boosted trees, Bayesian techniques, and nearest-neighbor approaches. Although artificial neural networks (ANNs) and DL might be used on data that is tabular in nature, alternative approaches frequently surpass neural networks with complete connectivity. The assessment of supervised ML models relies on datasets that contain both the input and the output. The method that is most prevalent is cross-validation, in which the set of data is divided into pairs in order of sets for training and testing that do not overlap. The model gets trained and optimized on the initial set of training for every split, whereas the predictive power of the trained model is evaluated on the test set. The average of all splits is the final score [11].

5.3 A comprehensive approach to pregnancy

Current developments in ML, such as multimodal learning, multiview representational learning, and multitask learning, offer an unprecedented chance for detailed modeling of pregnancy and its problems. These fields of study seek to create a cohesive framework (model) by combining datasets from numerous multiple domains and across various tasks (e.g., prediction of results or outcomes).

Multimodal models have already been used in the biological integration of data and are especially important for multiomics assessment as well as integration. At exactly the same time, the results of pregnancy are extremely linked and might point to distinct phenotypes with the same ailments.

This interconnectedness might be utilized via multitask learning that utilizes data included in associated results to strengthen models by favored solutions that have common biological structures throughout these results. DL has recently made significant contributions to the advancement of multitask learning [12]. The combination of multimodal learning and multitask methods allows for reduced descriptions of inputs and modeled phenotypes (results/outcomes) that lead to a novel comprehensive knowledge of pregnancy's underlying techniques or processes. Fig. 5.1 shows the factors associated with and the numerous information methods used to create a comprehensive approach or model for pregnancy assessment. Pregnancy can be impacted by a variety of factors that may be tracked and measured via different methods. This generates extensively multifaceted information that ranges from tabular data such as clinical indicators to far more intricate information such as ultrasonic images. Multimodal ML, which is primarily allowed by DL, is a subset of ML that aims at incorporating multiple data techniques into one predictive model.

This is accomplished by combining models that do well at interpreting all these techniques into one model. The incorporation of histology images and genetic expression statistics, for example, is being illustrated to enhance pan-cancer prediction. Simultaneously, the results of pregnancy are extremely interconnected and may indicate distinct phenotypes with the same type of disease. Multitask learning can take advantage of this interconnectedness. Rather than determining only one result, such as preeclampsia or preterm birth,

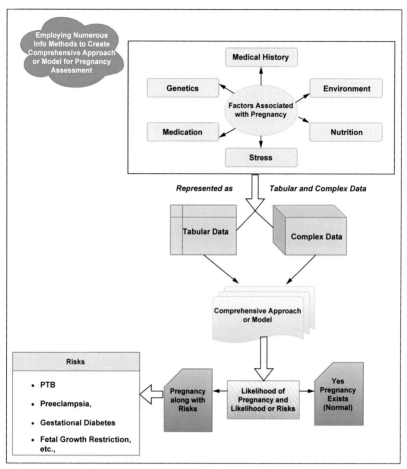

FIG. 5.1 Factors associated with and numerous information methods used to create a comprehensive approach or model for pregnancy assessment.

multitask models estimate multiple results at the same time [13]. To attain greater accuracy in prediction, the model must acquire the universal fundamental mechanisms of these results, if properly configured. This additionally enables us to exploit the information found within associated results and strengthen models by favored solutions that possess prevalent (e.g., biological) processes throughout these results; however, it may also offer novel insights into those prevalent procedures by comprehending the acquired interactions. Finally, multitask and multimodel models blend each of these ideas and use multiple data techniques for predicting multiple activities at the same time, and may open up novel opportunities for developing comprehensive models and improving one's comprehension of pregnancy [14].

There are several pregnancy assessment techniques that can be used in ML, including the following:

- Classification algorithms: These algorithms are a part of ML algorithms, used to classify pregnancy results. For example, the algorithm could be trained to classify pregnancies as successful or unsuccessful based on numerous factors, such as weight, maternal age, and medical history.
- Time-series analysis: This involves analyzing data over time, such as changes in hormone levels during pregnancy. ML algorithms can be trained to detect patterns and anomalies in time-series data to identify pregnancy-related events, such as ovulation or implantation.
- Natural language processing: This involves analyzing and understanding human language. In the context of pregnancy assessment, natural language processing can be used to analyze medical notes, patient records, or social media data to identify mentions of pregnancy and related issues.
- Medical imaging analysis: Medical imaging analysis involves analyzing medical images, such as ultrasounds or MRI scans, to detect pregnancy-related abnormalities or complications. ML algorithms can be trained to analyze these images and identify patterns that could indicate potential issues.

Following are seven generally used steps that can be taken to build an ML model [15,16] for early pregnancy prediction:

- Data collection: Collect data on various factors that can influence pregnancy, such as age, weight, menstrual cycle, and medical history. The data can be collected from medical records or through surveys.
- Data cleaning and preparation: Clean and preprocess the data to remove missing or erroneous values, normalize the data, and prepare it for use in ML algorithms.
- Feature selection: Choose the most relevant features to include in the ML model. This can be done using statistical analysis, feature importance metrics, or domain expertise.
- Model selection: Choose a suitable ML algorithm for the task, such as logistic regression, DTs, or SVMs.
- Training and validation: Train the ML model on a subset of the data and validate the model on a separate subset to evaluate its accuracy and generalizability.
- Optimization: Optimize the model parameters, such as the learning rate, regularization, or several hidden layers, to improve its performance.
- Deployment: Deploy the model in a clinical setting and monitor its performance over time.

The success of an ML model for early pregnancy prediction depends on the quality and quantity of the data used to train and validate the model. Additionally, the model must be validated in a real-world clinical setting to ensure its effectiveness and safety.

5.4   Significant clinical obstacles in pregnancy

There are significant clinical obstacles in pregnancy. Fig. 5.2 lists these obstacles, and we explain them in the section that follows:

- Pregnancy progression is dependent on complex biological changes: A healthy pregnancy is dependent on several interconnected biological changes. Abnormalities in maternal immune adaptation, placentation, and hormonal homeostasis, between additional processes, are involved in the pathogenesis of negative effects. For example, throughout pregnancy, the maternal immune system has to preserve tolerances to the fetoplacental unit in addition to protecting the fetus and mother from infectious agents. Consequently, the inability to perform has a direct connection to nearly all negative pregnancy results. Similarly, particular chronological changed within the maternal endocrine system orchestrate pregnancy phases, alongside any irregularities of hormone homeostasis that may give rise to pathogenesis [17].
- Pathogenic pregnancy processes cause complications: Pathogenic procedures can disrupt crucial biological changes throughout pregnancy, leading to complications during pregnancy. These procedures result in preeclampsia, preterm birth, and fetal growth restriction (FGR), which occurs in 15%–20% of pregnancies. While most of these illnesses manifest clinically in the third trimester of pregnancy, treatment following the first sign of clinical symptoms induces minor to slight enhancements. Because uterine remodeling, implantation, and placental growth occur at the start of pregnancy, determining molecular alterations weeks before clinical disease begins is an opportunity to significantly enhance detection and treatment while reducing mortality and morbidity among mothers and babies [18].
- Prematurity: This is still the primary cause of neonatal mortality and morbidity, and children born prematurely frequently experience perpetual obstacles, such as long-lasting physical and neurological impairments. Preterm birth etiologies are frequently classified into smaller groups.

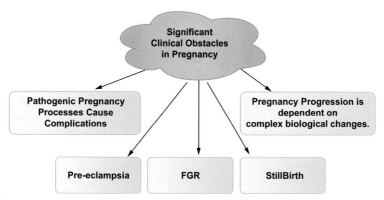

FIG. 5.2   Obstacles in pregnancy.

- Preeclampsia: This condition is a multiorgan illness that affects 2%–8% of pregnant women. Although sudden elevated blood pressure and proteinuria are the signs of preeclampsia, the condition can have serious consequences for several organs, the nervous system, lung function, liver function, the urinary tract, and heart. Preeclampsia can manifest as either early-onset or late-onset preeclampsia, with the latter type indicating worsening symptoms and a greater risk of adverse effects [18–20].
- Fetal growth restriction: FGR is frequently referred to as a determined fetal size that is below the tenth percentile; the majority of circumstances are "constitutional." Pathological FGR affects 3%–5% of pregnancies and is caused by infections (like cytomegalovirus), abnormal genetics, structural abnormalities, and/or underlying maternal problems that induce reduced perfusion of the placenta. While the majority of FGR cases are identified in the third trimester of pregnancy, determining FGR in either of the first two trimesters is difficult, as there are no established treatments to alter fetal development in utero [21].
- Stillbirth: This is defined as fetal demise soon after 20 weeks of pregnancy. Stillbirth is a worldwide medical burden in low- and middle-income countries (LMICs). The reasons for intrauterine neonatal death are intricate. Typical issues that can result in stillbirth involve fetal congenital defects, placental diseases, female reproductive disorders, and obstetric ailments. In addition, a stillbirth pregnancy will have mental and physical effects on the wellbeing of the mother and may jeopardize the viability of future gestation [22,23].

Apart from the preceding, there are several challenges that must be addressed when using ML for pregnancy assessment. Some of these challenges include the following [24,25]:

- Data quality: The accuracy and completeness of the data used to train ML algorithms are critical to the success of the system. Medical data can be highly variable and inconsistent, and data quality can be affected by factors such as missing data, errors, and bias [26].
- Limited data availability: In some cases, there may be limited data available to train ML models, which can make it difficult to achieve high accuracy. This is especially true for rare or complex conditions.
- Interpreting results: ML algorithms can produce highly accurate results, but interpreting those results can be challenging. It is important to ensure that the results are clinically relevant and understandable to healthcare professionals and patients [27,28].
- Ethical considerations: The use of ML in health care raises ethical questions, such as the potential for bias, discrimination, or infringement of privacy rights. These concerns must be carefully addressed to ensure that ML is used ethically and responsibly [29].

- Regulatory approval: ML models used in health care must be approved by regulatory authorities, such as the Food and Drug Administration in the United States before they can be used in clinical practice. The approval process can be time-consuming and expensive and requires rigorous testing and validation to ensure the safety and efficacy of the system.

Overall, while ML has the potential to revolutionize pregnancy assessment, it is important to address these challenges and ensure that the technology is used ethically and responsibly.

5.5 Conclusion

Early pregnancy assessment using ML has the potential to improve accuracy and efficiency in diagnosing and managing pregnancy-related conditions. With the use of ML algorithms, healthcare providers can analyze vast amounts of data collected from various sources, such as patient history, clinical observations, and laboratory test results, to make more informed decisions.

Some examples of how ML can aid in early pregnancy assessment include predicting the likelihood of miscarriage, identifying women at risk of preterm birth, and detecting fetal anomalies. ML algorithms can also help with personalized treatment planning and monitoring to improve pregnancy outcomes. However, it is important to note that the use of ML in early pregnancy assessment is still in the nascent stages, and more research is needed to validate the accuracy and reliability of these tools.

Additionally, ethical and privacy concerns related to the use of patient data in ML algorithms need to be addressed to ensure that these technologies are used in a responsible and ethical manner. Overall, the integration of ML into early pregnancy assessment has the potential to revolutionize the way healthcare providers diagnose and manage pregnancy-related conditions, leading to improved maternal and fetal outcomes.

References

[1] Z. Loring, S. Mehrotra, J.P. Piccini, Machine learning in 'big data': handle with care, EP Europace 21 (9) (2019) 1284–1285.

[2] A. Dhillon, A. Singh, Machine learning in healthcare data analysis: a survey, J. Biol. Today's World 8 (6) (2019) 1–10.

[3] M.W.L. Moreira, J.J.P.C. Rodrigues, F.H.C. Carvalho, N. Chilamkurti, J. Al-Muhtadi, V. Denisov, Biomedical data analytics in mobile-health environments for high-risk pregnancy outcome prediction, J. Ambient. Intell. Humaniz. Comput. 10 (10) (2019) 4121–4134.

[4] A. Sandstrom, J.M. Snowden, J. Hoijer, M. Bottai, A.K. Wikstrom, Clinical risk assessment in early pregnancy for preeclampsia in nulliparous women: a population-based cohort study, PLoS One 14 (2019) e0225716.

[5] N. Dehingia, A. Dixit, Y. Atmavilas, D. Chandurkar, K. Singh, J. Silverman, A. Raj, Unintended pregnancy and maternal health complications: a cross-sectional analysis of data from

rural Uttar Pradesh, India, BMC Pregnancy Childbirth 20 (1) (2020), https://doi.org/10.1186/s12884-020-2848-8.

[6] I.H. Sarker, Machine learning: algorithms, real-world applications, and research directions, SN Comput. Sci. 2 (3) (2021) 1–21.

[7] Maternal Mortality. https://www.who.int/publications/i/item/9789240068759. (Accessed 14 April 2023).

[8] D. Despotović, A. Zec, K. Mladenović, N. Radin, T.L. Turukalo, A machine learning approach for an early prediction of preterm delivery, in: 2018 IEEE 16th International Symposium on Intelligent Systems and Informatics (SISY), IEEE, Manhattan, 2018, pp. 000265–000270.

[9] M.N. Islam, T. Mahmud, N.I. Khan, S.N. Mustafina, A.N. Islam, Exploring machine learning algorithms to find the best features for predicting modes of childbirth, IEEE Access 9 (2020) 1680–1692.

[10] P. Cecula, Artificial intelligence: the current state of affairs for AI in pregnancy and labour, J. Gynecol. Obstet. Hum. Reprod. 50 (2021) 102048, https://doi.org/10.1016/j.jogoh.2020.102048.

[11] S. Ghaderighahfarokhi, J. Sadeghifar, M. Mozafari, A model to predict low birth weight infants and affecting factors using data mining techniques, J. Basic Res. Med. Sci. 5 (3) (2018) 1–8.

[12] C. Mangold, S. Zoretic, K. Thallapureddy, A. Moreira, K. Chorath, A. Moreira, Machine learning models for predicting neonatal mortality: a systematic review, Neonatology 118 (2021) 394–405, https://doi.org/10.1159/000516891.

[13] N. Zabari, Y. Kan-Tor, Z. Shoham, Y. Shofaro, D. Richter, I. Har-Vardi, A. Ben-Meir, N. Srebnik, A. Buxboim, Delineating the heterogeneity of preimplantation development via unsupervised clustering of embryo candidates for transfer using automated, accurate and standardized morphokinetic annotation, medRxiv (2022), https://doi.org/10.1101/2022.03.29.22273137.

[14] D.K. Stevenson, et al., Towards personalized medicine in maternal and child health: integrating biologic and social determinants, Pediatr. Res. (2020), https://doi.org/10.1038/s41390-020-0981-8 (Published online May 26).

[15] M. Lewandowska, B. Wieckowska, S. Sajdak, J. Lubinski, Pre-pregnancy obesity vs. other risk factors in probability models of preeclampsia and gestational hypertension, Nutrients 12 (2020) 2681.

[16] B.K. Sarkar, A two-step knowledge extraction framework for improving disease diagnosis, 8e Comput. J. 63 (3) (2020) 364–382.

[17] M.W.L. Moreira, J.J.P.C. Rodrigues, V. Furtado, C.X. Mavromoustakis, N. Kumar, I. Woungang, Fetal birth weight estimation in high-risk pregnancies through machine learning techniques, in: Proc. IEEE Int. Conf. Commun., May, 2019, pp. 1–6.

[18] I. Marin, N. Goga, Nutrition consultant based on machine learning for preeclampsia complications, Sci. Pap. D Anim. Sci. 62 (2) (2019) 82–87.

[19] I. Carrasco-Wong, M. Aguilera-Olguín, R. Escalona-Rivano, D.I. Chiarello, L.J. Barragán-Zúñiga, M. Sosa-Macías, et al., Syncytiotrophoblast stress in early onset preeclampsia: the issues perpetuating the syndrome, Placenta 113 (2021) 57–66, https://doi.org/10.1016/j.placenta.2021.05.002.

[20] P. Wadhwani, P.K. Saha, J.K. Kalra, S. Gainder, V. Sundaram, A study to compare maternal and perinatal outcome in early vs. late onset preeclampsia, Obstet. Gynecol. Sci. 63 (2020) 270–277.

[21] M. Lewandowska, B. Wieckowska, S. Sajdak, Pre-pregnancy obesity, excessive gestational weight gain, and the risk of pregnancy-induced hypertension and gestational diabetes mellitus, J. Clin. Med. 9 (2020) 1980.

[22] A. Khalil, et al., Change in the incidence of stillbirth and preterm delivery during the COVID-19 pandemic, JAMA 324 (2020) 705–706.

[23] B. Bekkar, et al., Association of air pollution and heat exposure with preterm birth, low birth weight, and stillbirth in the US: a systematic review, JAMA Netw. Open 3 (2020) e208243.

[24] H.K. Kang, A. Kaur, A. Dhiman, Menopause-specific quality of life of rural women, Indian J. Commun. Med. 46 (2) (2021) 273–276, https://doi.org/10.4103/ijcm.IJCM_665_20.

[25] H.K. Kang, A. Kaur, S. Saini, R. Ahmad, W.S. Kaur, Pregnancy-related health information-seeking behavior of rural women of selected villages of North India, Asian Women 38 (2) (2022) 45–64, https://doi.org/10.14431/aw.2022.6.38.2.45.

[26] R. Kaur, R. Kumar, M. Gupta, Food image-based nutritional management system to overcome polycystic ovary syndrome using deep learning: a systematic review, Int. J. Image Graphics (2022) 2350043.

[27] R. Kaur, R. Kumar, M. Gupta, Food image-based diet recommendation framework to overcome PCOS problem in women using deep convolutional neural network, Comput. Electr. Eng. 103 (2022) 108298.

[28] R. Kaur, R. Kumar, M. Gupta, Deep neural network for food image classification and nutrient identification: a systematic review, Rev. Endocr. Metab. Disord. (2023) 1–21.

[29] R. Kaur, R. Kumar, M. Gupta, Predicting risk of obesity and meal planning to reduce the obese in adulthood using artificial intelligence, Endocrine 78 (3) (2022) 458–469.

Chapter 6

Ensemble learning-based analysis of perinatal disorders in women

Malvika Gupta[a], Puneet Garg[b], and Chetan Malik[c]

[a]Department of CSE, ABES Engineering College, Ghaziabad, UP, India, [b]Computer Science and Engineering, St. Andrews Institute of Technology and Management, Farrukhnagar, Gurugram, Delhi NCR, India, [c]San Diego State University, San Diego, CA, United States

6.1 Introduction

In recent years, perinatal mental health has become a major focus in India, especially in rural regions. In a developing nation like India, it has become a matter of major concern. India's Government Health Organization is very concerned about mental health during pregnancy. Some pregnant women suffer from depression, which in turn can lead to suicidal thoughts, anxiety attacks, and sometimes miscarriage [1]. In the earlier 2000s, due to less awareness and monitoring, the mortality rate of women in India was high, especially in rural areas. However, since 2010, due to awareness programs organized by government health organizations, the mortality rate has reduced to more than 70%. The WHO has revealed data that shows that India has been successful in reducing maternal mortality rates significantly compared to the rates recorded in the early 2000s. This is great progress for any developing nation. According to the United Nations, the percentage of women in India suffering with pre and postnatal mental disorders is between 9% and 35%.

Fig. 6.1 describes some of the symptoms that may be present in women suffering from perinatal mental health disorders. These symptoms have been identified in case studies, medical research, and surveys conducted with the help of gynecologists. Results of the survey conclude that the period of pregnancy is the most critical period for a woman. During the study, researchers also discovered that perinatal mental health disorders can occur during both pre and post pregnancy. Postpregnancy symptoms can be diagnosed up to 1 year after giving birth. So, from the studies, it was identified that during pregnancy it can occur

Artificial Intelligence *and* Machine Learning *for* Women's Health Issues
https://doi.org/10.1016/B978-0-443-21889-7.00016-6

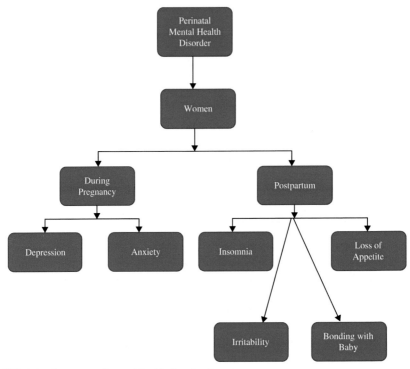

FIG. 6.1 Symptoms of mental health disorders in women.

two times either in the first trimester or the last trimester. Symptoms during this period are generally diagnosed as depression and anxiety. There is also a condition known as postpartum depression, which occurs after the birth of the child. This condition can occur up to 1 year after delivery. Symptoms of postpartum depression include the following:

- insomnia
- loss of appetite
- irritability
- lack of bonding with the babyMedical researchers suggest that these symptoms need to be diagnosed at a very early stage. For this, family members must cautiously monitor behavioral changes in the mother. Negligence might result in harm to the mother as well as the child.

Studies have shown that if a woman is suffering from a perinatal mental health disorder, this will likely affect the health of the fetus during the pregnancy and the health of the infant after delivery. Fig. 6.2 shows the issues that generally occur if a woman is suffering from a perinatal mental health disorder. During pregnancy, mental health issues in the mother could increase the chances of miscarriage and postpartum depression. There are also increased chances

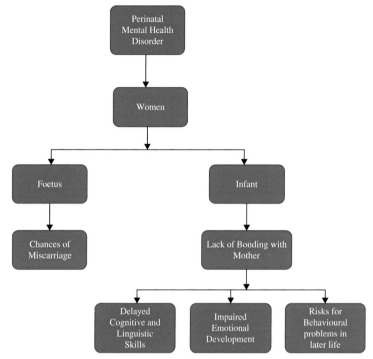

FIG. 6.2   Effects of perinatal mental health disorders on children before or after pregnancy.

of lack of bonding between mother and child due to which an infant may develop the risk of delayed cognitive and linguistic skills, lack of emotional development, and development of behavioral issues later in life. These issues have been identified via surveys and research, where women suffering from this disorder also committed suicide or ran away from their families.

The coronavirus pandemic has proven to negatively affect peoples' mental health. People have suffered from depression and anxiety attacks, according to reports from different health organizations. During the pandemic, pregnant women suffered severely from these symptoms, which led to suicidal thoughts and an increase in mortality rates. Nagendrappa et al. [2] and Jinhee Hur et al. [3] conducted a case study of mental health affects during the period of COVID-19.

During the COVID-19 pandemic, many people suffered from mental disorders for various reasons, including the following:

- loneliness
- lockdown
- infection with the virus and subsequent isolation from family
- financial losses
- fear of infection

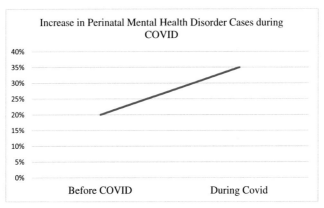

FIG. 6.3 Comparison of mental health disorders in pregnant women before and during COVID-19.

In studies conducted during the period of COVID-19, doctors suggested that pregnant women should not go out or be in contact with any person suffering from cough, cold, and other symptoms of the virus. Fig. 6.3 presents a graph that shows that 20% of women suffered from perinatal mental health disorders before COVID-19, and that this percentage increased by 9%–35% during the pandemic.

One study highlighted the concerns that women who were expecting or just gave birth reported to obstetricians during the COVID-19 pandemic. The majority of the 118 obstetricians who answered the online survey indicated that they had been called due to worries regarding hospital visits (72.6%), prevention strategies (60.17%), infant safety (52.14%), social media message anxiety (40.68%), and infection transmission (39.83%). Obstetricians believed that perinatal women with COVID-related anxiety needed access to services including movies, websites, and counseling [4]. Anxiety, depression, and high levels of stress have been demonstrated to be important impacts of pandemics, including COVID-19. Pregnant women are a susceptible group who may be concerned about the effects of COVID-19 infection on the fetus during pregnancy. There have been inconsistent findings about the morbidity and mortality of COVID-19 infection in pregnant women, including the potential for vertical transmission. Pregnant women have been reported to experience anxiety during infectious epidemics concerning a variety of aspects of childbirth, such as shattered expectations for prenatal and postnatal care. Natural disasters can harm the health of mothers and their newborns due to increased stress and disruptions in access to social support and care, according to a prior study on pregnant women. Women in the prenatal phase are under additional stress because of the COVID-19 pandemic, as symptoms of anxiety and sadness are made worse by social isolation and fear of infection [5].

Perinatal mental health disorders are being addressed more effectively with improvement in medical facilities, especially in urban areas, and thus these disorders are decreasing in incidence. However, in rural areas in India, these disorders persist because most people are not aware of or concerned about these issues [6]. According to reports from the WHO, however, India has successfully reduced its mortality rate in rural areas although there is still much work to be done [7,8]. All these surveys were conducted based on social demographic and health-related facilities on a particular population. By undertaking different case studies, it was decided that women should visit a psychiatrist in addition to her obstetrician during pregnancy [9].

One study observed that perinatal mental health disorders can lead to serious issues [10]. Perinatal mental health disorders in women are dangerous. Fig. 6.4 depicts the repercussions of these disorders, which generally occur during the first or last trimesters, as well as postdelivery. In both cases, mental health disorders may lead to depression, which can sometimes progress to. They can also

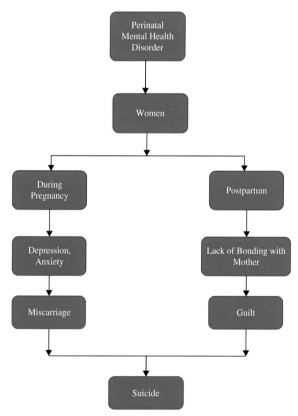

FIG. 6.4   Repercussions of perinatal mental health disorders in women.

affect the health of the infant, sometimes causing miscarriage or death of the infant after delivery, or other serious health issues.

Until the introduction of predictive modeling, or machine learning (ML) algorithms, perinatal mental health disorders were diagnosed after their peak. Thus, to avoid the severe repercussions of late diagnosis, it is important to diagnose mental health disorders in the early stages so that the proper treatment can be provided. For this, different types of applications have been developed to monitor blood pressure, heart rate, and so on, which in turn can be modeled using different ML techniques to predict whether a woman is suffering from a perinatal mental health disorder.

Fig. 6.5 shows the process followed by researchers for predicting perinatal mental health disorders in pregnant women using ML models. Table 6.1 compares prediction models like random forest (RF), Naïve Bayes (NB), and others based on different performance metrics, including area under the curve (AUC), F1-score, sensitivity, and specificity. These are performance metrics described in the classification report based on which the prediction can be easily categorized as up to what percentage the model on the dataset can give the best

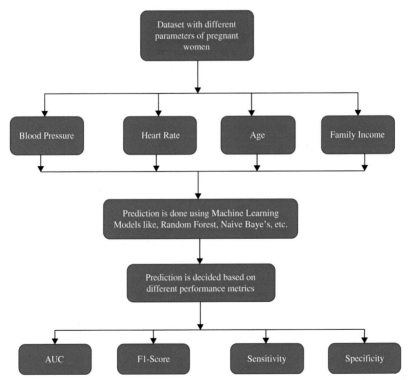

FIG. 6.5 Predicting perinatal mental health disorders in women using machine learning models according to different studies.

possible prediction. Nowadays, data is collected using different applications through smart devices like smartwatches. After collecting the data, data preprocessing is performed to remove null values and to identify that all the values under the important parameters should be present. After this, models are selected and applied to obtain results in the form of performance metrics. In research, it is usually expected that an ML model will provide true positive results. By true positive, the researchers will able to give a percentage that our model can predict this much percentage correctly or incorrectly.

Down To Earth is an app that was successfully utilized during the coronavirus pandemic in India. This chapter describes the different studies by government and private health organizations on perinatal health disorders in women in India. It also presents a survey of how effectively different types of predictive algorithms can be modeled. Table 6.1 describes the different models used for predicting perinatal health disorders.

Section 6.2 of the chapter includes the process of review, research questions, search criteria, and inclusion and exclusion criteria. Section 6.3 discusses the results and provides the answers to the research questions. Section 6.4 concludes the survey and highlights research gaps and future scope.

6.1.1   Process of review

We conducted our survey using 50 different articles dating from 2000 to 2023 (based on their access rights), including both general articles and articles based on ML prediction techniques. We also considered government studies. We identified that some government and private health organizations are using certain apps to monitor the mental health of women in India. Thus, we have included insights based on those applications in this research as well (Fig. 6.6).

Fig. 6.7 shows the number of papers selected based on the research domain. Articles were obtained from a variety of sources.

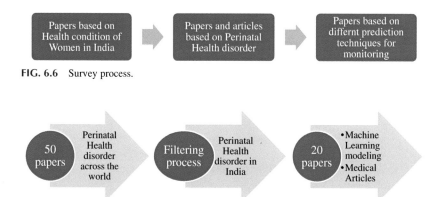

FIG. 6.6   Survey process.

FIG. 6.7   Papers selected based on relevance to the research.

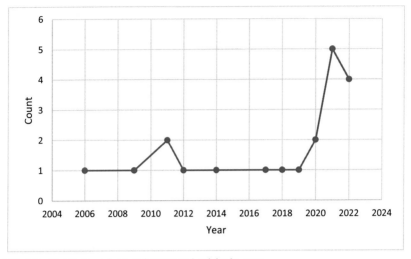

FIG. 6.8 Distribution of selected papers and articles by year.

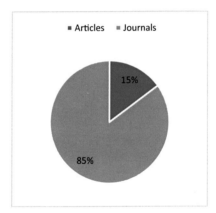

FIG. 6.9 Distribution of articles and journals.

Fig. 6.8 shows the yearly distribution of papers included in our survey.

Fig. 6.9 shows the percentage share of papers selected from different articles and journals.

6.1.2 Research questions

Q1. Is the mortality rate increasing due to perinatal mental health disorders of women during COVID-19 or not?

Q2. Can electronic gadgets or applications help diagnose this mental disorder?

Q3. Has the introduction of ML models helped in diagnosing perinatal mental health disorder at an early stage?

Q4. Which ML model is better for prediction? Also, what parameters are required to get better predictions?

6.1.3 Answers to research questions

A1. This question plays a very important role in analyzing the effectiveness of India's mental health care during the period of isolation or restriction. During COVID-19, apart from physical symptoms, mental health has become a matter of major concern for health organizations. This survey is based on the perinatal health of women; thus, we present two different case studies to help in understanding the concerns of increasing mortality rates during the pandemic.

Case 1: Before giving birth, a woman was diagnosed with COVID-19, and after delivering the baby she was not allowed to breastfeed, and she was kept away from the baby. Due to this, she started to experience depression, which lead to her neglecting the baby because of the guilt.

Case 2: Due to a perinatal mental health disorder, a young mother 25 years of age started exhibiting unwanted behavior towards family members and her infant. Her behavior escalated and the family had to isolate from her and provide her with psychological treatment.

A2. Yes, there are different types of awareness programs run by the government [15] that have helped in spreading awareness of perinatal mental health disorders. Electronic gadgets have helped in collecting data to help in predict these disorders. Different mobile apps have been developed to collect information, such as blood pressure, heart rate, body mass index, and more. This data is in turn fed to a suitable ML model to predict mental disorders at an early stage A [12].

A3. As the above research question provides half of the answer that data is collected through different types of smart gadgets and then after preprocessing the data, it is fed into the different models of ML for prediction. It might be possible that there will be no 100% accurate prediction but if an algorithm predicts that the vitals of a woman is showing that she is going to suffer from a perinatal health disorder. With the introduction of deep learning (DL) techniques may be with the help of gestures, it might give better results as compared to ML algorithms. The comparison between different ML models and DL is discussed in the answer to research question 4.

A4. See Table 6.1.

6.2 Literature survey

Fellmeth et al. [16] defined the mental health of pregnant women in India. This study focused on the repercussions of bad mental health during this period. It broadly suggests that women suffer from acute depression, which may lead

TABLE 6.1 Comparative analysis of different machine learning algorithms.

S. no.	Models used	Parameters for prediction	Performance
1	Random forest [11]	BMI, age, weight, height, smoking habits	73% accuracy
2	Machine learning and deep learning [12]	Survey data based on survey questions and audio recordings during pregnancy	Accuracy and F1-score
3	A deep neural network that uses a decision tree and support vector machine [13]	Diastolic and systolic blood pressure, heart rate, and age	98% accuracy
4	Support vector machine, Naïve Bayes, random forest, extreme gradient boosting, L2-regularized LR and decision tree [14]	256 variables are considered using different social-demographic conditions like age, BMI, blood pressure, heart rate, and so on.	Sensitivity, specificity, and AUC; 79% AUC for SVM, 98% sensitivity for decision tree, 61% specificity for Naïve Bayes
5	Boosted trees algorithm	BMI, heart rate, blood pressure, age	80% AUC
6	Classification tree algorithm	BMI, heart rate, blood pressure, age	84% accuracy, 73% AUC, 84% specificity, 73.4% sensitivity
7	Random forest, gradient boosting	BMI, heart rate, blood pressure, age, and dataset of COVID-19	80% AUC

to the death of the baby or the mother. Nagendrappa et al. [2] described the mental health conditions of women during the COVID pandemic. This study was carried out by focusing on two different case studies on how much COVID triggered mental health disorders in pregnant women.

Rathod et al. [7] described perinatal depression in rural India in a study based on the sociodemographic and health-related facilities in rural areas. The study focused on 226 and 130 women who were in the perinatal period. It concluded that women between the ages of 18 and 22 years have a greater chance of having perinatal mental health disorders. Koski et al. [17] described the cause of mental health disorders in women in rural areas. This study focused on domestic violence against women, due to which mental disorders can occur, leading to severe health deteriorating conditions and death. According to this study, many women in rural areas suffer from mental health disorders due to domestic violence.

Ghosh [15] described the use of electronic media to promote awareness of perinatal mental health. This study concluded that electronic media has helped a lot to spread awareness. Andersson ct al. [11] described the prediction of mental disorders based on different parameters using RF and lasso regression. This modeling achieved an accuracy of 73% in predicting whether a woman is going to suffer from depression.

Bilal et al. [12] described a smartphone application called Mom2B, which collects data based on survey questions and audio recordings during and after pregnancy. This data is then fed to a suitable ML method for prediction. As per the performance analysis, this study suggests that if there is audio data present DL may also be used for better performance. The prediction is analyzed based on the accuracy and F1-score of the performance metrics.

Khan et al. [18] surveyed the maternal health conditions in India and other developing conditions. The study surveyed the health-related facilities available in different countries. It also described how artificial intelligence (AI) can help to monitor the mental health status of women. Results showed that ML and DL techniques based on certain parameters can help easily predict perinatal mental health disorders in women.

Raza et al. [13] based their study on the idea that mental health is important for a healthy pregnancy. The study showed a prediction accuracy of 98% using the ensemble technique, which was a combination of neural network architecture with a DT and SVM. In the case of ML and DL, these algorithms generally give accurate results. This study used a dataset of major variables such as diastolic and systolic blood pressure, heart rate, and age. Wang et al. [14] described the prediction of perinatal mental health disorders based on different ML models, including SVM, NB, RF, extreme gradient boosting, L2-regularized LR, and DTs. The authenticity of prediction was decided based on different performance metrics, including sensitivity, specificity, and AUC. Results show a sensitivity of 98% result, indicating an accurate prediction. This study suggests that different ML algorithms give different values for these performance metrics. Thus, it is up to the researcher to choose the predictive model according to the best-suited performance metric.

Saqib et al. [19] defined the comparative analysis of how ML algorithms can help in predicting the mental health illness of a patient. This study was carried out for general mental health conditions, not for perinatal mental health disorders specifically. This survey suggest that ML models can help in the early prediction of perinatal mental health disorders in women. Betts et al. [1] suggest that the ML boosting trees algorithm outperforms logistic regression, achieving an AUC of 80%, which is one of the most important performance metrics.

This helps in the timely treatment of patients who show positive signs of suffering from depressive, bipolar, or perinatal mental health disorders.

Jinhee Hur et al. [3] described the prediction of women suffering from bipolar, depressive, or perinatal mental health disorders. The model used was classification trees, which achieved an accuracy of 84%. The study also calculated other performance metrics to obtain more accurate results. Qasrawi et al. [6] suggested RF and gradient boosting as the best-suited models for predicting mental health disorders in women, showing an AUC of 80%. Zhang et al. [8] studied postdelivery mental health disorders, to predict the disorder ML algorithm is used for early detection. The AUC score was 94% for this prediction.

6.3 Proposed methodology

Fig. 6.10 describes the proposed methodology for the prediction of perinatal mental health disorders. The dataset was collected via different apps and surveys conducted with the help of gynecologists. Not much research has been carried out on this topic using the concepts of ML and DL. Initially, the dataset can be collected from gynecologists, who can provide the details required for predicting perinatal mental health disorders during pregnancy, including in

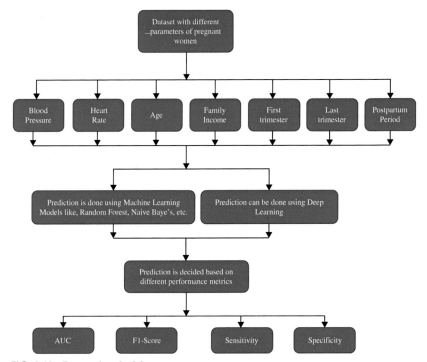

FIG. 6.10 Proposed methodology.

the first and last trimesters as well as the postpartum period. It is important to identify any issues at an early stage to avoid serious consequences for both women and infants.

6.4 Conclusion

In this chapter, we carried out a detailed survey of numerous medical articles focusing on perinatal mental health disorders in women in India. We presented a detailed review of the condition from the early 2000s up to today, and included articles on the mental health condition of women in rural areas to highlight the difference in the perinatal mental health of women in urban and rural areas. We found that the mental health of women is often ignored due to a variety of reasons. In India, during the study, many medical researchers concluded that due to societal barriers and other issues women are not able to express their problems, even among family members. Even if mental health issues are identified, society may not acknowledge them, and thus pregnant women may experience miscarriage or commit suicide. Many studies also illustrate that if a mother is not able to take care of the child after birth, she may be suffering from depression, which may lead to adverse consequences like running away from the family or suicide [20].

Our survey has shown the advantage of using different ML algorithms and DL for predicting mental health disorders. For this, we used difference performance metrics, including accuracy, sensitivity, specificity, F1-score, and AUC. Based on these metrics, it is easy for researchers to determine which ML algorithm or combination of ML algorithms is best suited to predict risk of mental health disorders in pregnant women. This survey concludes that different ML algorithms can be used for prediction at an early stage so that appropriate treatment can be provided as soon as possible.

6.5 Research gaps

During this study, we identified some research gaps. If these gaps are narrowed, India may be able to reduce incidence of depression among pregnant women even further. Some of these research gaps include necessity of ML and DL models for analysis, and lack of proper data, predictive analysis, and detection technology. In addition, many consulting doctors do not suggest psychological followup for their pregnant patients.

6.6 Future scope

After reviewing several ML algorithms, more algorithms can be used for prediction based on different datasets with more predictors. This prediction can be carried out on a rural dataset as still there are lots of scopes to improve the mental health condition of women. With the help of these ML algorithms, and based on their performances, an ensemble technique can be devised to obtain better results. If CT scans or X-rays are included as predictors, then DL is the best method for prediction. Including imaging studies may help in early-stage detection of mental health disorders, leading to earlier and hopefully more effective treatment.

References

[1] K.S. Betts, S. Kisely, R. Alati, Predicting postpartum psychiatric admission using a machine learning approach, J. Psychiatr. Res. 130 (July) (2020) 35–40, https://doi.org/10.1016/j.jpsychires.2020.07.002.

[2] S. Nagendrappa, et al., Perinatal mental health care for women with severe mental illness during the COVID-19 pandemic in India—challenges and potential solutions based on two case reports, Front. Glob. Women's Heal. 2 (July) (2021) 1–7, https://doi.org/10.3389/fgwh.2021.648429.

[3] E.G. Jinhee Hur, S.A. Smith-Warner, E.B. Rimm, W.C. Willett, W. Kana, Y. Cao, 乳鼠心肌提取 HHS public access, J. Int. Soc. Burn Inj. 43 (5) (2017) 909–932, https://doi.org/10.1097/MLR.0000000000001467.Using.

[4] M.H. Nanjundaswamy, et al., COVID-19-related anxiety and concerns expressed by pregnant and postpartum women—a survey among obstetricians, Arch. Womens. Ment. Health 23 (6) (2020) 787–790, https://doi.org/10.1007/s00737-020-01060-w.

[5] L.M. Osborne, M.C. Kimmel, P.J. Surkan, The crisis of perinatal mental health in the age of Covid-19, Matern. Child Health J. 25 (3) (2021) 349–352, https://doi.org/10.1007/s10995-020-03114-y.

[6] R. Qasrawi, et al., Machine learning techniques for predicting depression and anxiety in pregnant and postpartum women during the COVID-19 pandemic: a cross-sectional regional study, F1000Research 11 (2022), https://doi.org/10.12688/f1000research.110090.1.

[7] S.D. Rathod, S. Honikman, C. Hanlon, R. Shidhaye, Characteristics of perinatal depression in rural central, India: a cross-sectional study, Int. J. Ment. Health Syst. 12 (1) (2018) 1–8, https://doi.org/10.1186/s13033-018-0248-5.

[8] Y. Zhang, S. Wang, A. Hermann, R. Joly, J. Pathak, Development and validation of a machine learning algorithm for predicting the risk of postpartum depression among pregnant women, J. Affect. Disord. 279 (September) (2021) 1–8, https://doi.org/10.1016/j.jad.2020.09.113.

[9] I. Brockington, R. Butterworth, N. Glangeaud-Freudenthal, An international position paper on mother-infant (perinatal) mental health, with guidelines for clinical practice, Arch. Womens. Ment. Health 20 (1) (2017) 113–120, https://doi.org/10.1007/s00737-016-0684-7.

[10] C.H. Legare, et al., Perinatal risk and the cultural ecology of health in Bihar, India: perinatal health in Bihar, India, Philos. Trans. R. Soc. B Biol. Sci. 375 (1805) 2020, https://doi.org/10.1098/rstb.2019.0433.

[11] S. Andersson, D.R. Bathula, S.I. Iliadis, M. Walter, A. Skalkidou, Predicting women with depressive symptoms postpartum with machine learning methods, Sci. Rep. 11 (1) (2021) 1–15, https://doi.org/10.1038/s41598-021-86368-y.

[12] A.M. Bilal, et al., Predicting perinatal health outcomes using smartphone-based digital phenotyping and machine learning in a prospective Swedish cohort (Mom2B): study protocol, BMJ Open 12 (4) (2022), https://doi.org/10.1136/bmjopen-2021-059033.

[13] A. Raza, H.U.R. Siddiqui, K. Munir, M. Almutairi, F. Rustam, I. Ashraf, Ensemble learning-based feature engineering to analyze maternal health during pregnancy and health risk prediction, PLoS One 17 (11) (2022), https://doi.org/10.1371/journal.pone.0276525.

[14] S. Wang, J. Pathak, Y. Zhang, Using electronic health records and machine learning to predict postpartum depression, Stud. Health Technol. Inform. 264 (1) (2019) 888–892, https://doi.org/10.3233/SHTI190351.

[15] D. Ghosh, Effect of mothers' exposure to electronic mass media on knowledge and use of prenatal care services: a comparative analysis of Indian states, Prof. Geogr. 58 (3) (2006) 278–293, https://doi.org/10.1111/j.1467-9272.2006.00568.x.

[16] G. Fellmeth, et al., Perinatal mental health in India: protocol for a validation and cohort study, J. Public Heal. (U.K.) 43 (2021) II35–II42, https://doi.org/10.1093/pubmed/fdab162.

[17] A.D. Koski, R. Stephenson, M.R. Koenig, Physical violence by partner during pregnancy and use of prenatal care in rural India, J. Health Popul. Nutr. 29 (3) (2011) 245–254, https://doi.org/10.3329/jhpn.v29i3.7872.

[18] M. Khan, M. Khurshid, M. Vatsa, R. Singh, M. Duggal, K. Singh, On AI approaches for promoting maternal and neonatal health in low resource settings: a review, Front. Public Heal. 10 (September) (2022) 1–23, https://doi.org/10.3389/fpubh.2022.880034.

[19] K. Saqib, A.F. Khan, Z.A. Butt, Machine learning methods for predicting postpartum depression: scoping review, JMIR Ment. Heal. 8 (11) (2021) 1–14, https://doi.org/10.2196/29838.

[20] G. Fellmeth, S. Harrison, C. Opondo, M. Nair, J.J. Kurinczuk, F. Alderdice, Validated screening tools to identify common mental disorders in perinatal and postpartum women in India: a systematic review and meta-analysis, BMC Psychiatry 21 (1) (2021) 1–10, https://doi.org/10.1186/s12888-021-03190-6.

Chapter 7

Machine learning approaches to predict gestational diabetes in early pregnancy

Poonam Joshi[a], Sapna Rawat[b], Arpit Raj[c], and Vikash Jakhmola[a]
[a]Uttaranchal Institute of Pharmaceutical Sciences, Uttaranchal University, Dehradun, Uttarakhand, India, [b]JBIT Group of Institution, Dehradun, Uttarakhand, India, [c]Quantum University, Roorkee, India

7.1 Introduction

Gestational diabetes mellitus (GDM) is a common and alarming disease that is often seen in patients who are either planning for pregnancy or who are already pregnant. It typically develops around weeks 22–25 of pregnancy, the period in which the fetus develops the most [1]. Recent studies have found that GDM often arises due to a lack of awareness about diabetic conditions in pregnant women. Left untreated, GDM not only impacts women's health but also adversely affects the developing fetus, leading to neonatal obesity and cardiac-related issues. Statistics reveal that approximately one in seven fetuses is affected by these diseases [2,3].

Efforts to diagnose GDM effectively led to the establishment of organizations such as the National Institute of Diabetes and Digestive and Kidney Diseases (NIDDK). Their findings underscore the direct or indirect association between GDM and obesity or overweight conditions. The International Association of Diabetes and Pregnancy Study Groups (IADPSG) further contributes to understanding GDM causes, revealing that it often presents multiple risk factors. Notably, individuals with polycystic ovarian syndrome (PCOS) face a higher risk of GDM.

To address these challenges, medical science has witnessed innovations incorporating cutting-edge technologies. Specifically, artificial intelligence (AI)-powered equipment, utilizing advanced machine learning (ML) programs, plays a pivotal role in diagnosing GDM at its early stages. These technologies not only diagnose but also exhibit the capability to distinguish between GDM and nonGDM patients who may present with similar conditions but who differ in their etiological effects. This discrimination is crucial for effective diagnosis.

Artificial Intelligence *and* Machine Learning *for* Women's Health Issues
https://doi.org/10.1016/B978-0-443-21889-7.00011-7

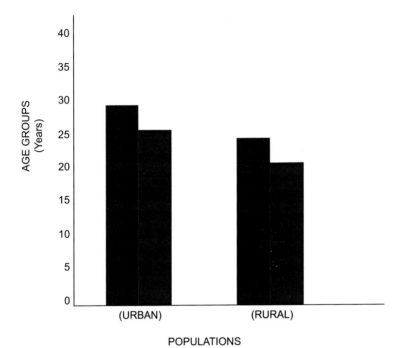

GRAPH 7.1 Comparison between GDM and nonGDM patients in urban and rural populations.

Interestingly, the prevalence of GDM is observed more commonly among urban women compared to their rural counterparts, as illustrated in Graph 7.1 [4,5]. This data underscores the need for targeted interventions and awareness campaigns in urban areas to mitigate the impact of GDM.

- GDM patients
- NonGDM patients

As this graph signifies, the occurrence of GDM and nonGDM in urban areas as compared to rural areas is due to unhealthy diet and lifestyle factors that must be maintained during pregnancy conditions and early diagnosis [6,7].

7.1.1 Detection techniques

There are numerous tests for diagnosing GDM. We discuss some of these in the sections that follow.

Glucose screening test

Early detection of glucose levels is crucial, and specific laboratory techniques have been established for this purpose. These techniques involve two distinct processes, as classified by global organizations [8].

Initially, the International Association of Diabetes and Pregnancy Study Groups employed a single-step approach that involved administering a specified amount of oral D glucose (approximately 75 g) and assessing plasma glucose levels. The results were categorized based on fasting (≥ 92 mg/dL), 1-h postadministration (≥ 180 mg/dL), and 2-h postadministration (≥ 153 mg/dL). If values exceeded these thresholds, GDM was confirmed.

Later advancements introduced a two-step approach governed by the American College of Obstetricians and Gynecologists and the National Institutes of Health. In the first step, 50 g of oral glucose was administered, and plasma glucose levels were checked at 1 h. The result was compared with the standard range of 135 mg/dL–140 mg/dL. Subsequently, in the second step, oral glucose was increased to 100 g and the results were evaluated after 3 h. The analysis involved comparing the data against the National Diabetes Data Group cutoff (fasting ≥ 105 mg/dL, 1-h postadministration ≥ 190 mg/dL, 2-h postadministration ≥ 165 mg/dL, 3-h postadministration ≥ 145 mg/dL) and the Carpenter and Coustan cutoff values (fasting ≥ 95 mg/dL, 1-h postadministration ≥ 180 mg/dL, 2-h postadministration ≥ 155 mg/dL, 3-h postadministration ≥ 140 mg/dL). If the obtained results exceeded these values, GDM was confirmed [9,10].

Oral glucose tolerance test (OGTT)

The oral glucose tolerance test (OGTT) stands out as a unique and fundamental laboratory test for detecting GDM. Typically conducted between the 24th and 28th weeks of gestation, this test follows a distinctive set of criteria. It involves the administration of approximately 75 g of glucose orally, specifically in the morning after an overnight fasting period of about 8 h.

The evaluation of results entails a comparison with standard values: fasting (≥ 5.1 mmol/L), one-h postadministration (≥ 10.0 mmol/L), and two-h postadministration (≥ 8.5 mmol/L) [11]. This test is a vital component in identifying GDM and contributes to comprehensive assessments during the crucial mid-pregnancy period.

Subcutaneous adipose tissue (SAT) measurement

This method involves measuring subcutaneous adipose tissue (SAT) depth via ultrasonography from the upper cutaneous layer of the abdominal region, especially the rectus-associated muscles. Some studies have shown a positive correlation between SAT depth and GDM [12].

Visceral adipose tissue (VAT) measurement

Visceral adipose tissue (VAT) measurement involves obtaining samples from specific muscle groups, commonly the abdominal region, with a focus on the rectus abdominis. This procedure entails measuring specific borders within the abdominal aortal region, allowing for precise delineation of the inner parts and their ranges to the anterior portion through the walls.

By strategically capturing images and creating a comprehensive report, medical practitioners gain a clear view of the internal adipose tissue. This innovative approach enables the early diagnosis of GDM during pregnancy. The incorporation of advanced imaging techniques, including zoom capabilities, enhances the accuracy and clarity of the obtained images [13,14].

Classification and regression trees (CART)

Classification and regression trees (CARTs) leverage specially designed ML programs capable of effectively incorporating extensive datasets from specific populations. Their primary aim is to enhance the treatment of insulin-related issues, particularly those associated with diagnostic results such as body mass index (BMI) and OGTT test outcomes.

The implementation of these advanced techniques involves linking ML programs with highly sophisticated algorithms. The synergy of these elements allows for the accurate screening and discrimination of GDM patients from nonGDM individuals. One prominent ML algorithm employed in this context is the eXtreme Gradient Boosting (XGBoost) approach, known for its effectiveness in obtaining precise results for early GDM detection [15].

AdaBoost

AdaBoost is considered to be the most innovative criteria for distinguishing GDM patients from nonGDM patients. This approach relies on cutting-edge ML programs that excel in evaluating the effectiveness of datasets sampled from affected patients. The methodology is centered on leveraging the frequency of interrelated datasets, where a higher frequency correlates with more effective results.

AdaBoost's strength lies in its ability to achieve greater accuracy, providing invaluable support in the early diagnosis of GDM. By harnessing advanced ML techniques, AdaBoost enhances the assessment of patient data, contributing to more accurate and timely identification of GDM.

Light gradient-boosting machine (LightGBM)

LightGBM is an advanced technique for detecting GDM during the early stages of pregnancy. This method utilizes a specific wavelength of light rays and operates based on a specially designed algorithm. It obtainins results in the form of a histogram, allowing for effective calculation and presentation of the obtained data.

The application of LightGBM enables medical practitioners to produce accurate results and provide enhanced treatment for GDM [16,17].

Extreme gradient boosting (XGB)

Extreme gradient boosting (XGB) involves evaluating many datasets and effectively discriminating relevant datasets from irrelevant ones. XGB achieves this by boosting the importance of relevant datasets and distinguishing them from the noise and irrelevant data encountered during the diagnostic process.

The multilayered result obtained through this helps in the effective early detection of GDM [18].

ML logistic regression

ML logistic obtains results from evaluating input data from affected patients. In this process, the input datasets are subjected to proper diagnosis, and the resulting output takes the form of specific signs and symbols. These symbols are assigned using specially designed coding, which enhances accuracy and ensures more precise interpretation of the results.

Decision trees (DTs)

Decision trees (DTs) are specially designed to individually evaluate input datasets by distinguishing them at several distinct levels, including classes and subclasses. DTs involve assessing all classes, obtaining information about GDM patients, and generating outputs that can distinguish GDM patients from nonGDM patients.

Random forests (RFs)

Random forests (RFs) are advanced algorithms based on the nonlinearity model and are characterized by frequent branching of classes. They work in combination with the results obtained from DTs. Unlike individual DTs, RFs compare results from multiple DTs simultaneously, without direct correlation with the results from linear-based decision algorithms. RFs effectively evaluate results individually, and the final outcomes provide valuable insights for discriminating between GDM patients and nonGDM patients [19].

Clinical decision support systems (CDSS)

Clinical decision support systems (CDSS) are platforms that establish effective connectivity between patients and medical practitioners, facilitated by the innovative mHealth application. This application is specifically designed to raise awareness about better treatments for patients with GDM. It plays a crucial role in providing accurate and effective decisions, enabling proper patient counseling during the diagnostic process.

mHealth addresses multifactorial aspects of GDM, including lifestyle maintenance issues, changes in intracellular glucose levels (a leading cause of early pregnancy GDM), and considerations for associated medications. This application acts as an adjunct to patient care, offering advice and preventive measures

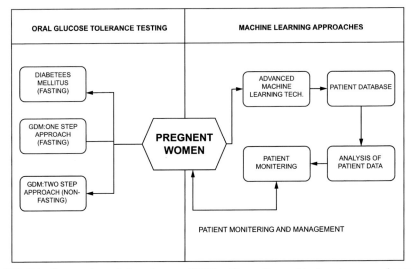

FIG. 7.1 Comparative techniques between OGTT and innovative machine learning approaches to detect GDM.

against GDM. By introducing specially designed apps, patients can receive guidance early in their pregnancy, enhancing their understanding of the condition and supporting preventive measures. Fig. 7.1 illustrates the comparative monitoring of GDM in pregnant women.

CDSS

The primary objectives of CDSS include creating a broader space for patients to discuss health concerns and enabling medical specialists to prevent the spread of diseases among healthy individuals. They provide innovative platforms for enhanced interaction between patients and medical specialists. Specially designed apps serve as conduits for communication, allowing individuals to discuss health issues with experts in specific fields.

CDSS operate on combinatorial systems, which involve knowledge-based systems. These systems systematically follow rules during programming, ensuring a methodical approach to decision-making. The knowledge-based aspect involves the utilization of expert rules in the design and execution of artificially created apps.

The knowledge-based component of CDSS involves the systematic application of rules during programming. This ensures that the app operates on a foundation of accumulated medical knowledge, contributing to informed decision-making.

CDSS also incorporate nonknowledge-based systems, which operate solely on input data. These systems use programmed languages to manipulate input data effectively, making decisions based on the provided information and generating results accordingly.

Fig. 7.2 represents CDSS functionality.

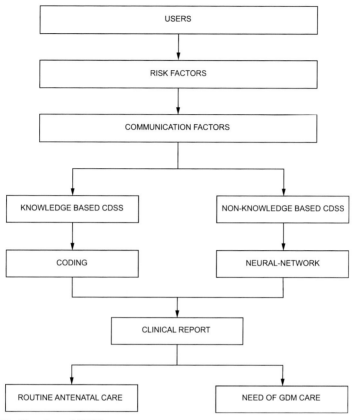

FIG. 7.2   Flowchart of CDSS.

CDSS methodology

These support systems have proven to be helpful for many patients during pregnancy. CDSS follow the steps listed here.

- Affected patients are advised to download specific apps to their smartphones.
- Patients are encouraged to keep their smartphones with them at all times. The apps are designed to automatically record various activities, including physical activity, changes in body weight, and any symptoms or history of diseases, especially those associated with GDM.
- Specially designed apps can effectively maintain records of patients over specific time periods. They automatically generate comprehensive reports for patients, which are shared on specific clinical linked portal systems. This allows medical practitioners to assess the current situation of patients and implement better management processes against GDM.

- After the initial evaluation of patient records, medical specialists conduct thorough reviews during appointments with affected patients. This revisiting of records by medical specialists ensures a comprehensive understanding of the patient's health status and aids in devising effective strategies for managing GDM.

LightGBM

LightGBM is an innovative approach aimed at reducing the prevalence of GDM among young pregnant women. This method has proven effective in studies, demonstrating a 20% reduction in the risk associated with GDM among affected pregnant women.

It operates on specially designed ML programs with the capability to classify GDM patients and nonGDM patients. This classification is crucial for effective diagnosis and subsequent intervention. It is adept at detecting GDM at the genetic level. In cases where patients have a familial history of diabetes, LightGBM reduces the impact of specific genes associated with GDM risk. It has been well evaluated that the role of LightGBM in diagnosing the GDM at its early stages had shown the better and preventive results. The results were shown by effectively reducing the count of specific genes, i.e. SNP 34 by 2 which was found to be 3 at its initial diagnosing procedure.

LightGBM outperforms other technologies like RF and XG Boost, achieving an accuracy of about 85.2%. This high level of accuracy is especially notable when compared using statistical measures such as receiver operating characteristic (ROC) curves [20], as shown in Fig. 7.3.

LightGBM methodology

The LightGBM methodology leverages the latest technology to produce effective results, combining two innovative approaches: gradient-based one-sided sampling system (GOSS) and the exclusive feature bonding algorithm (EFB). The following procedures outline the methodology:

- The algorithm-based technology collects all datasets related to particular patients.
- Then, the methodology effectively discriminates the collected datasets into distinct components. This segmentation is crucial for the targeted treatment of the disease.
- After obtaining individual patient data, the methodology employs the latest-based techniques to generate histograms. These histograms provide a visual representation of the data, aiding medical practitioners in easily classifying patients with GDM from those without GDM.

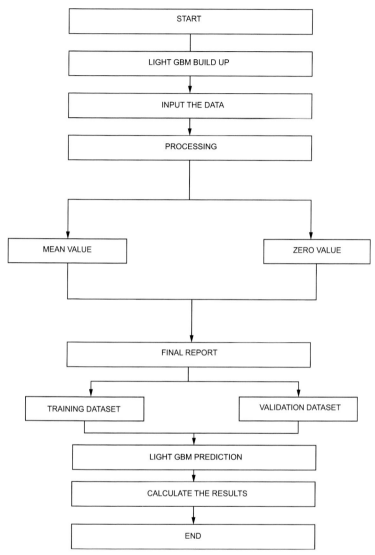

FIG. 7.3   Flow chart of LightGBM.

This unique process not only contributes to reducing risk factors associated with GDM but also ensures the accurate retrieval of results [21]. LightGBM achieved an accuracy of 85.2%, which is higher than that of CDSS.

7.2 Case study

Objective: To determine patient GDM status using LightGBM techniques.

7.2.1 Methodology

To study the effectiveness of LightGBM in determining GDM in pregnant women, we used a dataset comprising samples from various hospitals, labs, and testing agencies. The following procedures were employed to achieve high accuracy in GDM detection:

- LightGBM utilizes the latest-based histogram to distinguish datasets received from different sources into specified sections, ensuring that each section receives relevant information about the diseases.
- The input datasets are converted into suitable formats to facilitate effective result extraction.
- Distinct sections yield unique forms of histograms, which can be used to help diagnose GDM.
- The obtained results are classified based on gradients using the gradient-based one-side sampling (GOSS) technique. Gradients in trace amounts or lower amounts are detected, while higher amounts are left unchanged or stored for future use.
- Recent studies have shown that the obtained results include sparse features, representing both zero and nonzero values. The exclusive feature bundling (EFB) algorithm is employed to group these features, producing a new record of distinct values. This is crucial for accurate prediction of GDM at early pregnancy stages.

Table 7.1 summarizes the detection technique, accuracy, precision, and AUC of different methods. Graph 7.2 illustrates the difference in AUC among various detection techniques. These findings underscore the robustness and efficiency

TABLE 7.1 Observation table.

S.No.	Detection techniques	Accuracy	Precision	F1	AUC
1.	Logistic Regression	0.82356	0.58369	0.61485	0.87795
2.	LSTM	0.83415	0.64678	0.67885	0.92092
3.	LightGBM	0.85250	0.66341	0.69072	0.92104
4.	XGBoost	0.85770	0.75308	0.67847	0.91907

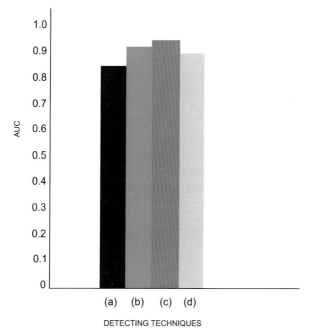

GRAPH 7.2   AUC vs Detection techniques.

of LightGBM in GDM detection, making it a valuable tool for healthcare professionals.

- Logistic Regression
- LSTM (Long Short-Term Memory)
- LightGBM
- XGBoost

7.2.2   Results

The statistical data show that LightGBM has an AUC of about 0.92 and an accuracy of 85.2%, suggesting it is better than other methods at detecting GDM in the early stages of pregnancy [22].

7.3   Future scope

Recent studies emphasize the significant benefits of introducing AI into the field of medical science, particularly for the early detection of diseases with multiple factors. GDM, being influenced by genetic and environmental factors, serves as a case in point. Establishing better platforms for the treatment of diseases at their primary stages is essential, and AI proves instrumental in achieving this goal.

Advanced technologies, including algorithms like XGBoost, LightGBM, and others, are specifically designated for early GDM detection. These algorithms work collaboratively, leveraging their capabilities to evaluate extensive datasets and discriminate between affected and nonaffected patients. The integration of AI not only facilitates early detection but also empowers patients with awareness of their health status.

Innovative techniques based on AI continue to evolve, focusing not only on disease presence detection but also on understanding the multiple factors that contribute to specific diseases. Specially designed health apps play a pivotal role in implementing these techniques. These applications assist in the early detection of diseases [22,23].

7.4 Conclusion

Insulin-related issues can lead to complications during pregnancy, commonly referred to as gestational diabetes mellitus (GDM). If left untreated, GDM can pose risks, particularly in young pregnant women, and often manifests around the 25th–26th week of pregnancy, which is a critical developmental period for the fetus.

Recent studies indicate that factors contributing to GDM include imbalanced diets, fluctuations in insulin levels, and various unhealthy conditions that directly or indirectly impact both fetal development and the health of pregnant individuals. To address these concerns, this chapter explored cutting-edge technologies aimed at reducing the risk of GDM during pregnancy.

One such technology is CDSS, offering a platform for direct communication with medical specialists. Additionally, we discussed smart applications, which monitor and manage patients' activities such as physical exercise and dietary habits. These apps also provide alerts for emergencies or critical conditions related to GDM.

Furthermore, the chapter discussed advanced algorithms like Random Forests, Decision Trees, and LightGBM. Notably, LightGBM stands out as an innovative technique that classifies input data and presents results through histogram algorithms with an impressive accuracy of approximately 85.2%. Such technologies empower medical practitioners to detect GDM early in gestational periods, contributing to more effective intervention strategies.

References

[1] N.E. Nora, A. Elsayed, F.M. El-Ghamry, Utilizing fog computing and explainable deep learning techniques for gestational diabetes prediction, Neural Comput. & Applic. (2022) 1–20.

[2] Y. Wang, et al., Plasma lipidomics in early pregnancy and risk of gestational diabetes mellitus: a prospective nested case-control study in Chinese women, Am. J. Clin. Nutr. 114 (5) (2021) 1763–1773.

[3] T. Sun, et al., Elevated first-trimester neutrophil count is closely associated with the development of maternal gestational diabetes mellitus and adverse pregnancy outcomes, Diabetes 69 (7) (2020) 1401–1410.

[4] B. Bhavadharini, et al., Prevalence of gestational diabetes mellitus in urban and rural Tamil Nadu using IADPSG and WHO 1999 criteria (WINGS 6), Clin. Diabetes Endocrinol. 2 (1) (2016) 8.

[5] S. Elnasr, H. Ammar, Ultrasound markers for prediction of gestational diabetes mellitus in early pregnancy in Egyptian women: observational study, J. Matern. Fetal Neonatal Med. 34 (19) (2021) 3120–3126.

[6] M. Khan, M. Khurshid, M. Vatsa, R. Singh, M. Duggal, K. Singh, On AI approaches for promoting maternal and neonatal health in low resource settings: a review, Front. Public Health 10 (2022) 880034.

[7] Y. Ye, Y. Xiong, Q. Zhou, J. Wu, X. Li, X. Xiao, Comparison of machine learning methods and conventional logistic regressions for predicting gestational diabetes using routine clinical data: a retrospective cohort study, J. Diabetes Res. 2020 (2020) 4168340.

[8] B.J. Daley, et al., mHealth apps for gestational diabetes mellitus that provide clinical decision support or artificial intelligence: a scoping review, Diabet. Med. 39 (1) (2022) e14735.

[9] J.W. Lum, et al., A real-world prospective study of the safety and effectiveness of the loop open source automated insulin delivery system, Diabetes Technol. Ther. 23 (5) (2021) 367–375.

[10] F. Hou, Z. Cheng, L. Kang, W. Zheng, Prediction of gestational diabetes based on LightGBM, in: Proceedings of the 2020 Conference on Artificial Intelligence and Healthcare, 2020.

[11] X. Lu, J. Wang, J. Cai, Z. Xing, J. Huang, Prediction of gestational diabetes and hypertension based on pregnancy examination data, J. Mech. Med. Biol. 22 (03) (2022).

[12] A.R. Yuhan, F.M. Rafferty, L. Mcauliffe, C. Wei, An explainable machine learning-based clinical decision support system for prediction of gestational diabetes mellitus, Sci. Rep. 12 (1) (2022).

[13] R. Jader, S. Aminifar, Predictive model for diagnosis of gestational diabetes in the Kurdistan region by a combination of clustering and classification algorithms: an ensemble approach, Appl. Comput. Intell. Soft Comput. 2022 (2022) 1–11.

[14] M. Valadan, Z. Bahramnezhad, F. Golshahi, E. Feizabad, The role of first-trimester HbA1c in the early detection of gestational diabetes, BMC Pregnancy Childbirth 22 (1) (2022) 71.

[15] J. Zhang, F. Wang, Prediction of gestational diabetes mellitus under cascade and ensemble learning algorithm, Comput. Intell. Neurosci. 2022 (2022) 3212738.

[16] B.J. Koos, J.A. Gornbein, Early pregnancy metabolites predict gestational diabetes mellitus: implications for fetal programming, Am. J. Obstet. Gynecol. 224 (2) (2021) 215.e1–215.e7.

[17] S. Jaiswal, P. Gupta, Ensemble approach: XGBoost, CATBoost, and LightGBM for diabetes mellitus risk prediction, in: 2022 Second International Conference on Computer Science, Engineering and Applications (ICCSEA), 2022.

[18] R. Liu, et al., Stacking ensemble method for gestational diabetes mellitus prediction in Chinese pregnant women: a prospective cohort study, J. Healthc. Eng. 2022 (2022) 8948082.

[19] L. Xiuxiu, G. Xing, W. Yan, Z. Yue, W. Yuzhu, H. Hongpu, Ideas on the construction of the telemedicine system for the gestational diabetes mellitus based on the clinical decision support system, in: 2021 International Conference on Public Health and Data Science (ICPHDS), 2021.

[20] R. Kaur, R. Kumar, M. Gupta, Food image-based nutritional management system to overcome polycystic ovary syndrome using DeepLearning: a systematic review, Int. J. Image Graph. (2022).

[21] R. Kaur, R. Kumar, M. Gupta, Predicting risk of obesity and meal planning to reduce the obese in adulthood using artificial intelligence, Endocrine 78 (3) (2022) 458–469.

[22] R. Kaur, R. Kumar, M. Gupta, Food image-based diet recommendation framework to overcome PCOS problem in women using deep convolutional neural network, Comput. Electr. Eng. 103 (108298) (2022) 108298.

[23] P.J. Liu, et al., The predictive ability of two triglyceride-associated indices for gestational diabetes mellitus and large for gestational age infant among Chinese pregnancies: a preliminary cohort study, Diabetes Metab. Syndr. Obes. 13 (2020) 2025–2035.

Chapter 8

Contribution of artificial intelligence to improving women's health in pregnancy

Gulafshan Parveen[a], Poonam Joshi[b], Yashika Uniyal[a], Haidar[a], and Sapna Rawat[c]

[a]Guru Nanak College of Pharmaceutical Sciences, Dehradun, Uttarakhand, India, [b]Uttaranchal Institute of Pharmaceutical Sciences, Uttaranchal University, Dehradun, Uttarakhand, India, [c]JBIT Group of Institution, Dehradun, Uttarakhand, India

8.1 Introduction

Artificial intelligence (AI) refers to using a machine to simulate intelligent behavior with little assistance from humans. AI is a component of the Fourth Industrial Revolution and holds great promise for resolving critical health issues. AI tools can be classified based on their utility and the scientific method used to develop them. Machine learning (ML) is the most widely used AI method in the development of most AI applications [1].

In the upcoming decades, health care will change, and recent commentary has addressed this potential transformation of the healthcare system as well as that of the public health system. AI is a general purpose technology (GPT). Its capabilities are defined as an employable core set. This is a fundamental set of talents that can be applied in a range of scenarios to carry out a number of jobs.

It has also proven beneficial in modeling healthcare data for improving clinical service for infection control responses, population management, and illness management, especially in the world's aging population. Artificial narrow intelligence (ANI) excels at completing specific, well-defined tasks with no opportunity for subjectivity. Most contemporary AI applications fall within this group.

AI-based improvements in robotic surgery, for example, demonstrate how to optimize minimally invasive precision surgery rather than traditional open surgical approaches. The objective of ANI is to equal human intellect and capabilities. This component of AI is being intensively examined and researched.

Artificial Intelligence *and* Machine Learning
for Women's Health Issues
https://doi.org/10.1016/B978-0-443-21889-7.00008-7

121

AI can aid in the transition from conventional knowledge to clinical practice by obtaining information from substantial data on evidence-based practice (EBP) and outputting smart answers in a fraction of the time required by traditional translational research paths [1].

Most healthcare disciplines, including radiology, neurology, orthopedics, surgery, and oncology, can benefit from AI. To promote AI's beneficial role in the future of the healthcare system, it needs to be implemented in research and then extrapolated to clinical practice. We believe AI can fundamentally alter the way health care is delivered, with effects that go beyond improving the efficacy and effectiveness of care [2].

AI is to present it in terms of levels: First: Simple control program, second: Machine learning, Third: Deep learning. Fig. 8.1 presents the different levels of AI control programs.

8.1.1 Machine learning

According to Samuel's 1959 definition, ML is a field of computer study that provides computers with the capability to learn without being clearly

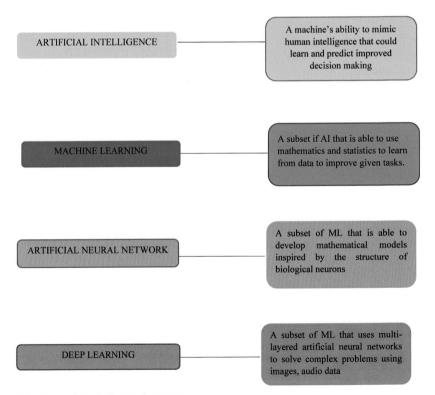

FIG. 8.1 Levels of AI control program.

programmed. According to Wikipedia, ML is the scientific study of algorithms and statistical models that computer systems use to perform individual tasks without explicit instructions. On sample data, ML is an algorithmic data-based model. This is referred to as training data. ML models are classified into several types, including artificial neural networks (ANNs), decision trees (DTs), support vector machines (SVMs), regression analysis, Bayesian networks (BNs), and genetic algorithms (GAs) [3].

AI is divided into two categories: ML techniques that analyze structured data and natural learning methods that extract information from unstructured data, such as clinical notes and medical papers, to complement and augment structured medical data.

AI can be used in four ways: to assess the probability of illness development and estimate treatment success, to alleviate or ameliorate medical issues, for patient care during the course of therapy, and to clarify pathology or mechanisms of disease to determine optimal treatment.

Many research studies have focused on predicting health outcomes in women, including those who suffer from diseases of the cardiovascular system, breasts, bones, cervix, and endometrium. This is a major focus of AI research in women's health. For example, the use of an ANN and a classification and regression tree to predict postmenopausal endometrial cancer in women has been studied [4].

AI has many stages for promoting maternal health (see Fig. 8.2).

8.1.2   Role of AI in pregnancy

The purpose of AI in pregnancy is to promote the mother's health at various stages of pregnancy, delivery, and postpartum (see Fig. 8.3). Maternal mortality

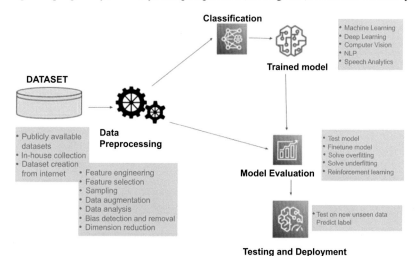

FIG. 8.2   Different stages in an artificial intelligence-based system.

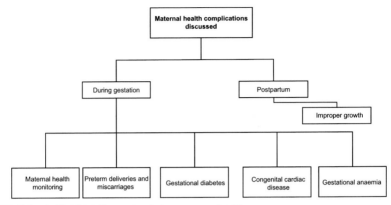

FIG. 8.3 Artificial intelligence's role in pregnancy health.

can be reduced through timely care of numerous maternal health concerns, such as premature births, miscarriages, gestational diabetes mellitus (GDM), heart disease, and postpartum depression.

8.1.3 Role of AI in neonatal health

AI also benefits newborn health. It contributes to decreasing the child mortality rate as well as the severity of the consequences if missing unprocessed (see Fig. 8.4). For example, AI can be used in newborns to evaluate pain and predict sepsis and malnutrition due to jaundice [5].

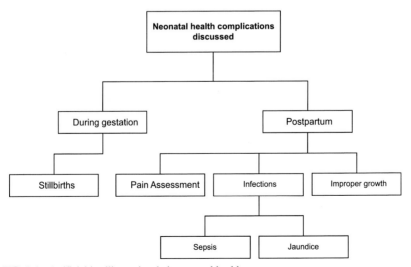

FIG. 8.4 Artificial intelligence's role in neonatal health.

Challenges of AI In women's health.

Figs. 8.5 and 8.6 highlight some of the challenges of using AI in women's health.

1. **Data availability and quality**

    Data availability and quality are the biggest challenges in AI because AI depends on having a large amount of data to train its models. In India, Internet connectivity is an issue, and thus, it is difficult to access data on women's health and childbirth. What data does exist is of poor quality, adding to the challenges of using AI effectively.

2. **Limited infrastructure**

    Limited infrastructure is another challenge. In India, problems with infrastructures such as electricity and internet connectivity make it difficult to use AI.

3. **Ethical concerns**

    Ethical concerns are important challenges for AI-based solutions. Some of these include regulation, legislation, safety, and privacy concerns.

4. **Language and dialects**

    Language and dialect barriers are significant challenges in India, which make it difficult to develop AI-based solutions. It can develop accurate and reliable. Which need a different community's language?

5. **Socioeconomic status**

    It can be quite difficult for those who are poor to access the technical services offered by AI solutions. Addressing socioeconomic disparities can help improve maternal-child health in India [6].

Solutions examples according to challenges (AI been adopted)

AI methods are clearly defined and differentiated for different stages in both the pregnant person and the fetus (see Fig. 8.7). AI and ML can be applied in

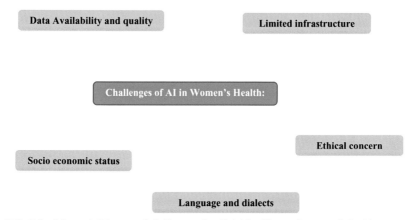

FIG. 8.5 Schematic Diagram of challenges of artificial intelligence in women's health.

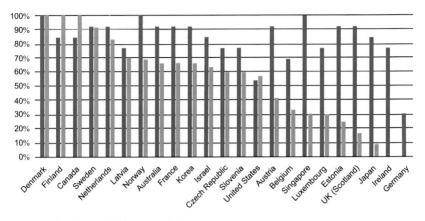

■ % of key national datasets available

■ % of available datasets regularly linked for research, statistics and/or monitoring (indicators)

FIG. 8.6 Schematic diagram of challenges in artificial intelligence systems.

different stages as well as areas, including preconception, the use of AI reproductive technologies, prenatal care, early pregnancy, premature birth, neonatal care, gynecological care, delivery and labor. To provide information on which stages of pregnancy are most studied, we examine the distribution of research by stage, making it easier to define gaps in the literature [7].

Using AI to obtain fetal images or videos has great potential to improve outcomes of artificial reproductive technologies (ARTs). For example, a convolutional neural network (CNN) was implemented to select the highest-quality embryos from a high-volume fertility center in the United States.

8.1.4 Medical imaging in pregnancy

(1) Fetal development

Both ultrasound and magnetic resonance imaging (MRI) are safe imaging procedures for expectant mothers, but they should only be performed when they are likely to provide patients with medical benefits or answers to pertinent clinical questions.

With very few exceptions, radiation exposure through radiography, computed tomography (CT), or nuclear medicine imaging techniques is at a far lower dose than exposure associated with harm to the fetus. A pregnant patient should not be denied access to these procedures if they are necessary in addition to ultrasonography or MRI for the diagnosis in question or if they are more accessible. Employing virtual organ computer-aided analysis (VOCAL), texture analysis, and CNN to enhance images of fetal organ development. Three papers examined CNN and texture analysis. Employing virtual organ computer-aided analysis (VOCAL), texture

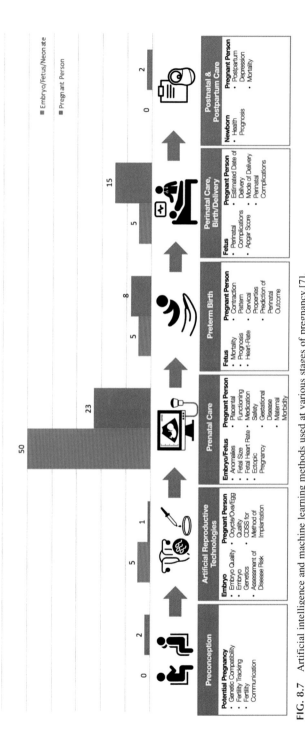

FIG. 8.7 Artificial intelligence and machine learning methods used at various stages of pregnancy.[7].

analysis, and CNN to enhance images of fetal organ development. Three papers examine CNN and texture analysis. Three other studies examined electronic health record (EHR) data using the high-performance ML approaches of DF and SVM.

(2) **Placental functioning**

The clinical methods that are now available to predict and evaluate placental health are still insensitive. The dynamic, noninvasive, real-time assessment of placental health and the early diagnosis of illnesses with a placental basis show considerable potential for advanced MRI methods. Based on the visible light spectrum of the placenta before birth, two kinds of placental imaging studies that are now available may be distinguished. The majority of predelivery placental imaging research has utilized pictures captured using MRI and ultrasound.

(3) **Genetics screening and chromosomal abnormalities**

Genetic screening can help in early detection of genetic conditions. First trimester screening makes use of maternal blood tests and fetal ultrasonography. The chance of a fetus developing a birth abnormality may be assessed with the use of this screening method. Multiple indicators, or parent screening, may involve many blood tests throughout the second trimester. These markers indicate the likelihood of having a baby with a specific genetic condition or birth defect. Ultrasounds may be performed at various points during pregnancy to assess fetal growth, estimate due date, and check for structural abnormalities in the body.

(4) **Gestational diabetes**

Insulin resistance, a condition that causes cells to utilize insulin less efficiently and increases the body's requirement for insulin, is a result of the body producing more hormones during pregnancy. All pregnant women suffer some insulin resistance in the later stages of their pregnancies. Yet, some women already have insulin resistance before being pregnant. They have higher insulin needs at the start of pregnancy and are more prone to develop GDM. GDM is a hyperglycemic condition that affects the health of both the mother and the fetus in the short and long term. Nevertheless, by the time a diagnosis is made, the fetal phenotype has already changed. GDM is often discovered between the 24th and 28th week of pregnancy. ML-based models have shown promise in predicting this condition, but they must be validated in different contexts and populations before they can be implemented in routine clinical practice.

(5) **Hypertension**

AI has been shown to be capable of identifying risk factors and phenotypes of hypertension (HTN), predicting the risk of incident HTN, diagnosing HTN, estimating blood pressure (BP), developing novel cuffless methods for BP measurement, and comprehensively identifying factors associated with treatment adherence and success. This strategy was chosen due to theoretical concerns for the welfare of the fetus stemming from reduced uteroplacental perfusion and in utero exposure to antihypertensive medicine, as well as the debate over whether pregnant women benefit from BP therapy normalization [7].

It is necessary to introduce education on AI into the women health

AI plays a significant role in the educational system, specifically in medical practitioner education. In the not-too-distant future, medical students will gain from AI-powered computer vision and mixed-reality solutions, which will provide an incredible learning experience and increase student interest in ML.

Medical professionals and students must be safeguarded and sufficiently knowledgeable to be able to interpret the outcomes of AI-based systems. Any medical professional using AI must have the skills to explain and clarify its use. Patient data can also be kept confidential and secure via AI-based technologies [8].

8.2 Future scope of AI in pregnancy

8.2.1 AI in the first trimester of pregnancy

The greatest risk of developing problems during pregnancy occurs during the first 3 weeks of fetal growth and development. According to first assessments, the first structure detected by two-dimensional ultrasonography (2DUS) is the gestational sac. There was a restriction placed on single section measurements by 2DUS, which may mean that there was no discernible difference between the gestational sac of normal and abnormal fetuses in the first trimester. The degree of fetal development can be further revealed by comparing 2D and 3D picture volume measurements.

For this purpose, a nuchal translucency/nasal bone (NT/NB) scan 3D-image volume calculation method has been introduced. This method is used to examine fetal cardiac activity and abnormality. It can also measure delivery date, image the cervical limb and placental lining, detect fetal age, measure brain development, and detect chromosomal disorders. It also performs fetal segmentation, standard biometry views estimation, automatic biometry measurements, and detection of fetal limbs. In one study, 104 fetal measurements and fetal appendages in the first trimester (10–14 weeks) were detected using AI to examine the anatomical components such as the fetus, gestational sac, and placenta via a 3D CNN algorithm. The NT should be assessed using a conventional median sagittal picture that indicates the maximal thickness between the fetal skin and the subcutaneous soft tissue at the cervical spine level.

8.2.2 AI-aided ultrasound in the second and last trimesters

Prenatal ultrasonography is employed to monitor fetal growth and development, find anomalies, and treat prenatal illnesses. One of the most challenging organs to evaluate with prenatal ultrasonography is the fetal brain. AI-aided ultrasonography in the developing brain can help to determine fetal head biometry, measure cranial capacity, and automatically segment fetal head circumference (HC) as well as the fetal head and its internal structure. With aberrant readings suggesting poor development, HC is a crucial indicator for diagnosing restricted growth. Measure the gestational age as well as the bi-parietal diameter (BPD), which is the width of the developing baby's skull from one parietal to the next (GA).

Craniocerebral development may be carefully assessed during pregnancy using a variety of measures and sections, and intracranial anomalies can be discovered. Most ultrasonic machines on the market today include semiautomatic measurement software that requires the localization of two points (typically for the short diameter locating point on the section of BPD), which could result in measuring error. To extract HC, AI additionally employs a CAD system based on the RF algorithm.

8.2.3 Automatic recognition and auxiliary assessment of fetal facial structures

With the assistance of skilled sonographers, AI-assisted 3D ultrasound also evaluates fetal facial anatomy as an additional measurement. Due to how long it takes to arrange the limbs or umbilical cord correctly, this approach is time consuming. Although 2D imaging can quantify some types of facial malformations, it cannot identify all deformities. Thus, AI plays a supporting role in prenatal facial recognition, craniofacial development, and the diagnosis of congenital abnormalities (see Fig. 8.8).

Preprocessing of ultrasound images by AI can enhance the effectiveness of diagnosis. For instance, Smart Face ultrasound may, with a single click, remove facial obstructions and improve visual angles while also automatically identifying key facial characteristics in 3D image and measurement data. Add many frames to the image of the fetal head, use AI to identify facial edges, distinguish facial contours, and remove facial occlusion to provide a 3D representation of the face's contours. The orientation of the face might also be determined using the fetal face model. For further in-depth analysis, it might be rotated to a specified angle.

8.2.4 Automatic recognition of the fetal brain for the diagnosis and prognosis of related diseases

Aiding in diagnosis and therapy is the ultimate objective of AI in medicine. The distribution of images and image recognition are efficient ways to improve the accuracy and dependability of ultrasonography. The fetal craniocerebral volume (comprising the cerebellum, brain, and frontal area) was measured, and a 3D picture of the unborn brain was made applying artificial boundaries to

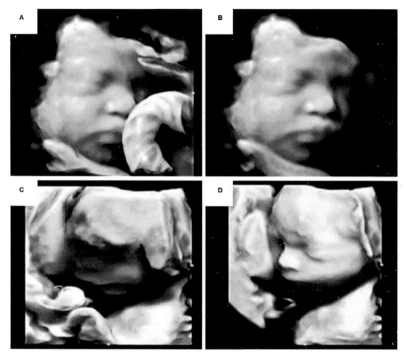

FIG. 8.8   3D ultrasound imaging of fetus with umbilical cord.

diagnosis poor growth. Ultrasound revealed that the craniocerebral volume of unusually tiny fetuses was smaller than that of normal fetuses.

Recently, there has been an increase in the application of AI in the diagnosis of fetal craniocerebral abnormalities. Partial corpus callosum agenesis, Dandy-Walker malformation (DWM), and other cerebellum-related illnesses may be identified by AI systems using fetal craniocerebral measurements and recognition data.

8.2.5   Accurate recognition and measurement of fetal abdominal structures

Abdominal circumference (AC) is the most accurate predictor of fetal weight. Standard abdominal sections can be used to assess both AC and fetal growth and development. The AC is more difficult to assess than the HC due to the low contrast of the fetal abdomen in acoustic imaging and the incidence of irregular abdominal echoes. However, in AI-based applications, the identification of the abdomen and measurement of the AC were comparable to those of the head; the main differences were the anatomical structures to be identified (i.e., the presence, position, direction, and distance of nearby organs such as the spinal cord, gallbladder, and stomach), as well as the effect of the amniotic fluid sac on boundary recognition. Transfer learning decreased model accuracy overfitting, and created a deep learning (DL)-based recognition system to standardize AC measurements.

Together with measuring fetal AC, the model was also capable of automatically arranging typical abdominal parts in the dynamic ultrasound picture. Using this information, another study separated the procedure into three stages: initial AC estimate, measurement, and plane acceptability verification. To increase measurement and location precision, a CNN and a U-Net were employed throughout each step. Several algorithm improvements and data volume amplifications have been used to address image recognition and autonomous measuring problems. Nevertheless, 3D reconstruction can help in determining the volume of the hollow viscera because 2D image measurements of the long diameter and area could not identify the viscera. As a result, the fetal bladder and stomach are presently measured using this technique.

Formal guidelines and regulation

Formal regulations and guidelines regarding use of AI in pregnancy may vary depending on the country or region. In general, uses and development of AI applications in health care, particularly during pregnancy, may be overseen by regulatory agencies such as the US Food and Drug Administration (FDA) or the European Medicines Agency (EMA). The EMA published a reflection paper on the application of AI in medicine. The regulatory environment for the creation and implementation of AI applications in health care, especially those employed during pregnancy, is discussed in this study.

The World Health Organization (WHO) and the International Society of Obstetrics and Gynecology (ISUOG) are two more organizations that may potentially offer advice or guidelines on the usage of AI apps during pregnancy.

The safety and efficacy of AI applications, ethical issues surrounding the use of AI in health care, and possible effects on patient outcomes and healthcare delivery are all things that regulatory and guideline governments may take into account. When thinking about using AI apps during pregnancy, it is critical for healthcare providers and patients to be aware of these rules and laws.

US Food and Drug Administration (FDA): The FDA regulates medical devices, including AI applications used during pregnancy. Any AI application intended to be used during pregnancy as a medical device must go through a premarket review process to ensure its safety and effectiveness.

European Commission: The European Commission released guidelines for trustworthy AI that outline principles and requirements for the creation and implementation of AI systems. These recommendations adhere to values like responsibility, justice, and transparency.

National Institute for Health and Care Excellence (NICE): In the United Kingdom, NICE provides guidelines for the management of pregnancy and childbirth. These guidelines include recommendations for the use of technology, including AI applications, to support women during pregnancy and childbirth.

World Health Organization (WHO): The WHO has published a report on the use of digital health technologies, including AI, to improve maternal and child health. The report provides guidance for the development and deployment of digital health technologies, including ethical considerations and recommendations for monitoring and evaluation.

Regulations and guidelines related to AI in pregnancy may still be evolving, as AI technology continues to advance and new use cases are developed. Women who are pregnant or considering pregnancy should always consult with their healthcare providers to determine the best course of action for their individual situation.

The WHO does not have specific guidelines for AI, but it has released reports and guidance related to the use of digital health technologies, which can include AI.

For example, in 2019, the WHO released a guideline on digital health interventions for health systems strengthening, which includes considerations for the design, implementation, and evaluation of digital health interventions, including those that use AI. WHO reports in this area include the following:

- Global Strategy and Action Plan on Aging and Health: The report emphasizes the importance of using AI and digital technologies to support healthy aging and promote wellbeing in older populations.
- Ethics and Governance of AI for Health: The paper offers a framework for the ethical development and application of AI in health care, with a particular emphasis on topics like transparency, accountability, and fairness.
- Digital Health: a call for government leadership and cooperation on information and communication technologies in the health sector: This report seeks to integrate digital technology and AI into medical systems so that they can deliver better results and support universal access to care.

Additionally, the WHO has collaborated with other organizations to develop ethical principles for AI, such as the Montreal Declaration for Responsible AI and the AI Ethics Guidelines for the European Commission. These principles focus on ensuring that AI is developed and used in ways that are transparent, accountable, and respectful of human rights.

IEEE Global Initiative on Ethics of Autonomous and Intelligent Systems: A thorough set of ethical standards for AI has been developed by the IEEE Global Initiative. These recommendations include a wide variety of topics, such as accountability, privacy, transparency, and the societal effects of AI.

European Union's Ethics Guidelines for Trustworthy AI: To encourage the creation and application of reliable AI, the European Union has also created a set of recommendations. The seven main aspects that are the emphasis of these standards include human agency and supervision, technical robustness and safety, privacy, and data governance, among others.

Sustainable Development Goals (SDGs): The United Nations set SDGs with the intention of promoting sustainable development and addressing global issues including poverty, climate change, and inequality. The advancement and application of AI must be in line with social responsibility and sustainability ideals if it is to help accomplish these objectives.

The Asilomar AI Principles: In 2017, a group of the world's leading AI researchers and experts developed the principles for Asilomar. These principles set out 23 recommendations on how to develop AI with safety and ethics, which cover topics such as transparency, fairness, and privacy.

The Montreal Declaration for Responsible AI: The Montreal Declaration is a set of principles developed in 2018 by researchers and practitioners working on AI. The Declaration highlights the need to ensure that AI is developed and used responsibly, with a view to ensuring its transparency, accountability, and protection of fundamental rights [9].

Regulation of AI: There are currently no specific regulations for AI in pregnancy, but the use of AI in health care is subject to regulations and guidelines related to medical devices, patient privacy, and ethics.

Medical devices that use AI for pregnancy monitoring or diagnosis are subject to regulatory oversight in many countries, such as the FDA and the European Union's CE marking process. These regulations ensure that medical devices are safe and effective for their intended use and undergo rigorous testing and evaluation before they are allowed on the market.

In addition to regulations for medical devices, the use of AI in pregnancy must also comply with patient privacy laws, such as HIPAA in the United States and GDPR in the European Union. AI systems must protect patient data and maintain confidentiality, ensuring that personal health information is not shared or misused.

Finally, the ethical use of AI in pregnancy should also be considered. AI systems should be transparent and accountable, and their use should be carefully monitored to ensure that they do not perpetuate bias or discrimination against certain groups of patients. Healthcare providers should also consider the potential impact of AI on patient-provider relationships and the need for informed consent when using AI in pregnancy care.

- Food and Drug Administration. The FDA regulates medical equipment in the United States, including those that use AI. Any AI-based medical device intended for use during pregnancy would need to undergo rigorous testing and approval processes before being authorized for use.
- In Vitro Diagnostics Regulation (IVDR) and Medical Devices Regulation in the EU. According to these laws, medical devices must go through a conformity assessment before being released onto the market [10].

In addition to regulatory oversight, ethical considerations should be taken into account when developing and deploying AI in pregnancy-related care. This includes ensuring that AI systems are transparent, explainable, and unbiased, and that they are used to supplement rather than replace human decision-making.

8.3 Conclusion

It is concluded that although AI has limitations and challenges, such as the necessity for large amounts of data and privacy and ethics concerns, it use in health care will continue to grow; thus, education about AI is important for all healthcare practitioners, including those practicing in maternal health. AI systems can improve pregnancy outcomes and patient care via early detection of maternal and fetal diseases and conditions. For example, the combination of AI with ultrasonography is helping physicians detect a variety of disorders and diseases, increasing productivity, decreasing missed and incorrect diagnoses, and significantly increasing the standard of medical care. The use of AI in obstetrics and gynecology has advanced significantly, although more research is still needed to determine the applicability and efficacy of many models.

References

[1] G. Delanerolle, Artificial intelligence: a rapid case for advancement in the personalization of gynecology/obstetric and mental health care, AI and Women's, Health (2021).

[2] A. Israa, Contribution of artificial intelligence in pregnancy: a scoping review, Inform, Technol. Clin. Care Publ. Health (2022) 333–335.

[3] T. Yoldemir, Artificial intelligence and women's health, Climacteric 23 (1) (2020) 1–2.

[4] J.M. Roberts, D. Heider, L. Bergman, K.L. Thornburg, Vision for improving pregnancy health: innovation and the future of pregnancy research, Reprod. Sci. 29 (10) (2022) 2908–2920.

[5] J. GeumHee, Artificial intelligence, machine learning, and deep learning in women's health nursing, Korean J. Women Health Nurs. (2020).

[6] R. Wang, R. Wang, et al., Artificial intelligence in reproductive medicine, Reproduction (2019).

[7] S. Chakraborty, Using AI to Improve Maternal and Child Health in India, the Integration of AI with the Already Existing Healthcare Systems Could Help Bring About a Substantial Difference in Maternal and Child Health in India, 2023.

[8] S. Secinaro, D. Calandra, A. Secinaro, V. Muthurangu, P. Biancone, The role of artificial intelligence in healthcare: a structured literature review, BMC Med. Inform. Decis. Mak. 21 (1) (2021) 125.

[9] A. Leimanis, K. Palkova, Ethical guidelines for artificial intelligence in healthcare from the sustainable development perspective, Eur. J. Sustain. Dev. 10 (1) (2021) 90.

[10] H. Stephen, et al., Ethics and Governance of Artificial Intelligence for Health WHO Guidance, WHO Guidance, 2021.

Chapter 9

Artificial intelligence-based prediction of health risks among women during menopause

Medha Malik[a], Puneet Garg[b], and Chetan Malik[c]
[a]*Department of CSE, ABES Engineering College, Ghaziabad, UP, India,* [b]*Computer Science and Engineering, St. Andrews Institute of Technology and Management, Farrukhnagar, Gurugram, Delhi NCR, India,* [c]*San Diego State University, San Diego, CA, United States*

9.1 Introduction

Menopause is a natural physiological process that causes reproductive difficulty in women. It encompasses a variety of physical and emotional symptoms that can affect wellbeing. In India, menopausal women are at increased risk of diseases, including cardiovascular disease (CVD), osteoporosis, breast cancer, and depression. The World Health Organization (WHO) has highlighted the health risks faced by menopausal women in India over the years. Menopausal women in India, according to the WHO, are at a greater risk of acquiring noncommunicable illnesses [1]. These problems occur due to lifestyle factors such as reduced physical activity, poor diet, and use of tobacco [2].

Fig. 9.1 describes the health issues that women face after menopause. These issues have been identified and confirmed via medical research papers, articles, and case studies based on different national and international data. Menopause can affect body mass index (BMI), mental health, and the cardiovascular system, and can cause cancer, which we discuss later in the chapter [3].

Menopause has become a major topic of concern in the rural part of India because of social barriers. Most of the women in the rural regions of the country do not have access to proper medical assistance. In addition, due to the pressures of society, women in rural India suffer from major health issues. Taking all of this into consideration, the WHO recommends that women going through menopause should get proper guidance and medical health checkups (e.g., blood pressure and cholesterol monitoring). The WHO also recommends different health organizations conduct health awareness programs, especially in rural India. These health awareness programs should focus on the repercussions of menopause [4].

Artificial Intelligence and Machine Learning for Women's Health Issues
https://doi.org/10.1016/B978-0-443-21889-7.00010-5

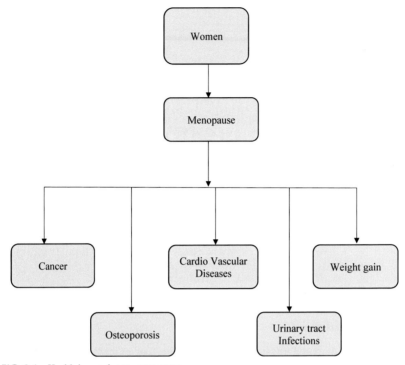

FIG. 9.1   Health issues due to menopause.

A recent research article on menopausal women in rural Bathinda, Punjab, examined the sociodemographic profile and health issues of postmenopausal women living in the region. The research found that most postmenopausal women in the area were in the age range of 50–59 years, had low levels of education, and were unemployed or engaged in unskilled labor. The most common problems faced by the women were fatigue (70.4%), backache (69.4%), breathing problems (52.2%), and abdominal discomfort (43.1%). The study emphasizes the need for a comprehensive approach to women's health that takes into account sociodemographic factors and cultural beliefs [5]. Another study investigated the efficacy of anthropometric elements for assessing the likelihood of osteoporosis in active, community-dwelling postmenopausal women across the countryside in southern India. The study found that BMI, waist circumference (WC), and waist-to-height ratio (WHtR) were significantly associated with risk of osteoporosis, with WHtR showing the strongest association. The study also found that a combination of WHtR and age can predict the risk of osteoporosis with greater accuracy than using either indicator alone. The study finds that anthropometric markers, notably WHtR, can predict the risk of osteoporosis in postmenopausal women and aid in the early detection and prevention of the illness [6].

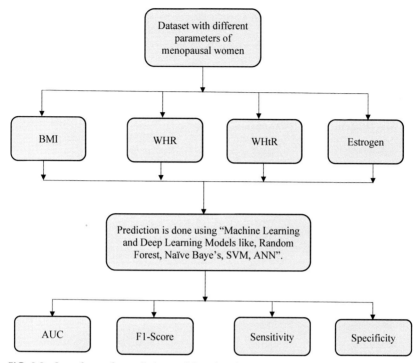

FIG. 9.2 Input factors for predictive models and performance metrices.

Fig. 9.2 depicts the range of factors that can be used for predicting various health risks. Many tools are being developed that can successfully identify the risk of certain health issues in menopausal women. These health issues include cervical cancer, weak bone density leading to osteoporosis, heart complications, and so on. The models proposed include various statistical models, such as machine learning (ML) and deep learning (DL) models like recurrent neural networks (RNNs), artificial neural networks (ANNs), support vector machines (SVMs), and others [7]. Many of these models are helpful in early detection and timely preventive measures for major health issues. Various factors are taken into consideration when developing a model for prediction in menopausal women (see Fig. 9.2) [8,9]. The performance metrics include factors such as area under curve (AUC), F1-score, sensitivity, and specificity. The model performs predictive analysis based on these performance metrics [10].

9.1.1 Research questions

Q1. Is menopause in women the cause of osteoporosis?
Q2. Is ovarian cancer risk higher during postmenopause or during menopause?
Q3. Are there clinical tools that can detect ovarian and cervical cancer?

Answers to research Questions:

Answer 1. Menopause is not the sole cause of osteoporosis in women, but it is a significant contributing factor. Osteoporosis is a disease that causes weak and brittle bones, which can lead to fractures and other health consequences.

During menopause, women experience a significant decrease in the production of estrogen, a hormone that helps to maintain bone density. This hormonal change can lead to an acceleration of bone loss, which can result in osteoporosis. However, there are other factors that can contribute to osteoporosis in women, including genetics, lifestyle factors such as smoking and consumption of alcohol, and medications [11,12].

Answer 2. Ovarian cancer is typically more prevalent in older women. Women older than 50 years account for the vast majority of ovarian cancer cases. According to one medical study, menopause is not a risk factor for ovarian cancer, but the hormonal changes that occur during and after menopause may have a role in the disease's development.

Research has shown that the risk of ovarian cancer may be greater in the years immediately following menopause, possibly due to a higher level of circulating hormones such as estrogen and testosterone. However, the risk of ovarian cancer continues to increase as women age, and postmenopausal women are still at a incomparably greater risk than premenopausal women [13–15].

Answer 3. For cervical cancer, a Pap smear is a common screening tool that can detect abnormal cells in the cervix before they become cancerous. A Pap smear is a procedure in which a healthcare professional removes cells from the cervix and sends them to a lab for examination. If abnormal cells are detected, further testing, such as a colposcopy, may be recommended.

In addition to Pap smears, there is also a test called the human papillomavirus (HPV) test, which can detect the presence of HPV, a virus that can cause cervical cancer. The HPV test may be used in conjunction with a Pap smear or as a standalone test [16].

There is no good screening test for ovarian cancer that can detect the illness in its early stages. However, some doctors may offer a blood test called the CA-125, which evaluates the levels of a protein that can be high in ovarian cancer patients. Imaging tests, such as transvaginal ultrasound, may also be used to help detect ovarian cancer [17,18].

9.1.2 Selection of articles

Methods

We conducted a bibliographic search in open-source databases such as PubMed, IEEE Explorer, National Library of Medicine, Scopus, Web of Sciences, and Science Direct using the phrases "menopause and artificial intelligence," "DL," and "health risks during menopause using DL." We removed studies with names that were not directly relevant to the research, as well as older data that showed repeated techniques. We chose papers dated 2010–23; however, we did find a few papers dated earlier that were relevant, and thus, we included them in this study.

Discussion

According to the collective research, there are numerous artificial intelligence (AI) techniques that can be used to improve risk assessment in women before and after menopause. This chapter collects and discusses the primary methodologies used to identify various health concerns in menopausal women while also recognizing their problems and trends. Techniques for categorizing risk groups that used AI principles, complex algorithms, and input parameters as major considerations are emphasized.

Selection of studies

To investigate the use of AI in menopause, we identified articles of various types, including systematic, randomized multicenter studies; cohort investigations; case sequences; and others.

1. Papers including parameters for diagnostics factors.
2. AI based data classification models.

Our survey was carried out using 56 different articles dating from 2010 to 2023. We included different articles and research papers in general as well as those based on some ML prediction techniques. We also included government studies of rural areas in India that reviewed the health risks in menopausal women. Few of the effective tools were proposed during the period 2010–23, which are effective as risk calculators, for predicting the possibilities of numerous conditions like fracture prediction, cancer prediction, etc. [5,19].

Fig. 9.3 depicts an organization chart of the technique used to choose the 23 papers for this study. The first step involved searching key phrases to identify relevant articles. Titles that corresponded with the survey's emphasis on the incorporation of AI in conjunction with diagnostic imaging and metrics as risk factors were carefully prescreened. Following that, we eliminated papers with repeating topics and few citations. To select which papers would be read completely, we excluded those that did not offer techniques related to the estimation of health risks in menopausal women.

9.2 Literature survey

Many tools have been developed for bone mineral density (BMD) measurement in postmenopausal women for osteoporosis risk assessment [20,21]. A new ML tool, the osteoporosis self-assessment tool (OST), aims at accurately identifying the risks of osteoporosis was developed using an SVM [22].

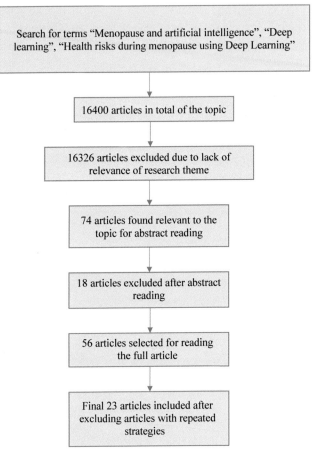

FIG. 9.3 Selection of studies.

The Breast and Ovarian Analysis of Disease Incidence and Carrier Estimation Algorithm (BOADICEA) is a risk model that incorporates the impact of
risk factors such as polygenic risk scores (PRS) and other risk factors (RFs)
in breast and ovarian cancer [23,24]. Yet another breast cancer prevalence prediction study discussed how, when all factors have been taken into consideration, the expected lifetime hazards for women in the UK population range
from 2.8% for the first percentile to 30.6% for the 99th percentile, with
14.7% of women anticipated to have a lifetime risk of 17%–30% (average risk
according to the National Institute for Health and Care Excellence [NICE] standards) and 1.1% of women to have a lifetime risk of 30% (high risk) [25].

A 2015 study based on a survey by the National Health Ministry of India
aimed to obtain information about the major risk of osteoporosis in women after
menopause [26].

In 2019, one study investigated basal metabolic rate (BMR) and its association with 13 cancers, including postmenopausal breast cancer, to identify

subgroups with greater risk of these health issues [27,28]. Using a statistical approach, the study found the specific major cancer risks in women were breast cancer, endometrial cancer, and ovarian cancer.

Considering the various tools available for osteoporosis risk evaluation, a comparative study of the various tools was conducted in Malaysia. The 2019 study compared results of six tools, based on 224 participants, who underwent the procedures. Of the total 224 participants, 164 showed the absence or presence of osteoporosis [29]. The precision of all tools was compared to and the best prediction was done by the Simple Calculated Osteoporosis Risk Estimation (SCORE) developed in the United States.

A 2020 study on application of ML approaches for predicting osteoporosis risk included K-nearest neighbor (KNN), decision tree (DT), random forest (RF), gradient boosting machine (GBM), SVM, ANN, and logistic regression (LR). The ANN model predictions were more accurate compared to other methods [30]. These prediction models could help in early prediction and prevention of osteoporosis.

A 2020 study investigated the causes of osteoporosis in postmenopausal women and found that menopause causes a deficiency of calcium in the body, which in turn leads to decreased bone strength [31].

A study conducted in 2022 investigated ovarian cancer, which is a cancer with a high mortality rate due to difficulties in diagnosis and improper screening [32]. This study identified a combination of raised carbohydrate antigen 125 (CA125) and transvaginal ultrasonography as the best method to detect ovarian cancer.

Another study investigated the use of colposcopy in the detection of cervical cancer, which generally has a tendency to occur in postmenopausal women [33]. Colposcopy is a successful screening tool, and 19 organizations have used this tool to detect cervical cancer and level of malignance.

In 2023, research was undertaken to construct an ML model to investigate and forecast the recurrence risk of breast cancer patients [34]. The training cohort included 1289 individuals and the test cohort included 552 patients. A total of 1841 textual reports from 2011 to 2019 were included. Long short-term memory (LSTM), XGBoost, and SVM showed recurrence risk prediction accuracies of 0.89, 0.86, and 0.78, respectively.

Another study examined a predictive tool called BFH-OST for osteoporosis in women after menopause [35]. This study was conducted in China, where it was observed that by using this predictive screening tool women older than 45 years suffered with osteoporosis. This study compared the screening tool with OSTA and found that BFH-OST performed better than OSTA with a sensitivity of 73% and specificity of 72%.

9.3 Comparison

Based on our study, there are many tools that are being developed for fruitful prediction and identification of various health concerns in women during and after menopause (Table 9.1).

TABLE 9.1 Comparison based on different models.

S.no.	Models	Motive	Observation	Results
1	LSTM, XGBoost, and SVM [34] 2020	Prediction and recurrence risk of breast cancer	The following factors were considered: distant organ metastasis, lymph node metastasis (including the number of lymph node metastases), HER-2, ER, PR, and Ki-67 expression; pathology grade; menopausal status; age; and lympho-vascular invasion	0.89, 0.86, and 0.78
2	FRAX [36] 2020	Fracture risk assessment	Age, gender, weight, height, prior fracture, parental fracture history, rheumatoid arthritis, usage of glucocorticoids, secondary osteoporosis, smoking, and alcohol intake are all factors considered	It was not that successful in diagnosing the fracture properly
3	BOADICEA [25] 2019	Breast and Ovarian Analysis of Disease Incidence and Carrier Estimation Algorithm (BOADICEA) risk model to incorporate the effects of polygenic risk scores (PRS) and other risk factors (RFs)	All genetic and lifestyle/hormonal/reproductive/anthropomorphic factors are considered jointly	The model should be suitable for counseling women consistently across different levels of clinical care
4	ANN [30] 2020	Prediction of osteoporosis risks	Various factors were taken into consideration, including height, BMI, history of smoking, history of fracture, duration of menopause, diabetes	ANN model with accuracy of 74%

5	SVM [22] 2013	The osteoporosis self-assessment tool (OST)	Factors included for assessment were BMI, duration of menopause, duration of breastfeeding, estrogen therapy, and hypertension, among others	Accuracy of 76.7%
6	SCORE [37] [38] [39] [40], 1998, 2001, 2002	"Simple Calculated Osteoporosis Risk Estimation tool"	Data of 1279 postmenopausal women. Factors included were age, weight, race, fracture history, rheumatoid arthritis history, and estrogen use	When combined with physical assessment, the tool gave efficient results of prediction
7	BFH-OST [35]	Screening tool for predicting osteoporosis	Data of 1721 women in China taking the following parameters into consideration: age, weight, height, and body mass index	It could be a cost-effective tool, but it had actually increased the risk osteoporosis
8	ORAI [41]	Osteoporosis Risk Assessment Instrument (ORAI)	Study of 1376 women in Canada, based on age, weight, and estrogen use	Effective in decreasing the need for DXA testing required for osteoporosis testing

9.4 Conclusion

Our research findings suggest that menopausal women are at an increased risk for several health issues, including CVD, osteoporosis, and depression. This risk can be mitigated through lifestyle modifications such as regular exercise, balanced diet, and hormone replacement therapy. However, it is critical for women to collaborate closely with their healthcare professionals to establish the best course of action for their specific requirements. More study is required to completely comprehend the intricacies of menopause health and to create effective preventative and treatment measures. Overall, our findings stress the significance of treating menopausal health as a vital aspect of overall wellbeing.

Fig. 9.4 shows the concerning rate of cases of various health risks in women with menarche. Considering the various factors, women with early menarche are at a greater risk of cancer, whereas the risk of cancer at later menopause age is at a higher risk as well [42]. The rates of ovarian cancer are greater in women during menopause, making early and proper treatment a major concern [43].

Fig. 9.5 shows the trends of health risks in menopausal women as they age [43]. Many tools are being proposed and clinically implemented to better predict these concerns.

9.5 Future scope

The use of AI in analyzing health risks in menopausal women is an emerging area of research with several potential future scopes, some of which include:

1. Predictive analysis: One potential future scope of research is the development of predictive models that use AI to identify menopausal women at high risk of developing certain health conditions, such as osteoporosis, CVD, or breast cancer. This could help healthcare providers tailor preventive

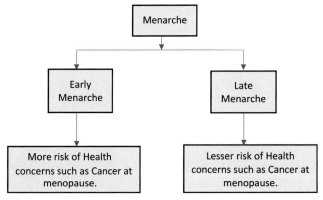

FIG. 9.4 Risk rate considering menarche.

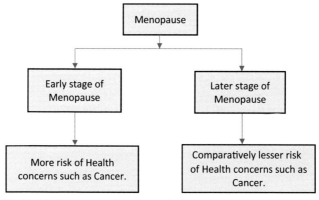

FIG. 9.5   Health risks in menopausal women as they age.

measures and interventions to individual women based on their unique risk profiles.

2. Personalized medicine: Another potential area of research is the use of AI to develop personalized treatment plans for menopausal women based on their individual health profiles. For example, AI could be used to analyze a woman's genetic data, medical history, and lifestyle factors to determine the best course of action for managing her menopause symptoms and reducing her risk of developing certain health conditions.

3. Remote monitoring: AI could also be used to develop remote monitoring systems that allow healthcare providers to track the health of menopausal women from a distance. For example, AI-powered wearables could be used to monitor a woman's heart rate, blood pressure, and other vital signs, alerting healthcare providers to any abnormalities or changes that may require intervention.

4. Improved diagnosis: AI could also be used to improve the accuracy and speed of diagnosis of certain health conditions in menopausal women. For example, AI-powered imaging technologies could be used to identify breast cancer at an earlier stage, when it is more treatable.

Overall, the use of AI in analyzing health risks in menopausal women has enormous potential to improve the quality of care for women during this important life stage. Further research is needed to fully explore the possibilities of this emerging field.

References

[1] A. Misra, et al., Nutrition transition in India: secular trends in dietary intake and their relationship to diet-related non-communicable diseases, J. Diabetes 3 (4) (2011) 278–292.

[2] W. H. Organization, OBESITY: Preventing and Managing the Global Epidemic, WHO Consultation, Geneva, 2004.

[3] G.M.V. Dijk, M. Kavousi, J. Troup, O.H. Franco, Health issues for menopausal women: the top 11 conditions have common solutions author links open overlay panel, Maturitas 80 (1) (2014) 24–30.

[4] R. Reid, B.L. Abramson, J. Blake, S. Desindes, S. Dodin, S. Johnston, T. Rowe, N. Sodhi, P. Wilks, W. Wolfman, Managing menopause, J. Obstet. Gynaecol. Can. 36 (9) (2014) 830–833.

[5] R. Samtani, D. Garg, N. Sharma, R. Deb, Sociodemographic pattern of postmenopausal women and health issues: a study in rural Bathinda, Punjab, J. Midlife Health 11 (3) (2020) 168–170.

[6] K. Sridharan, et al., Utility of anthropometric indicators in predicting osteoporosis in ambulant community dwelling rural postmenopausal women from southern India, Trop. Dr. 50 (3) (2020) 228–232.

[7] K. Zhao, H.-C. So, Drug repositioning for schizophrenia and depression/anxiety disorders: a machine learning approach leveraging expression data, IEEE J. Biomed. Health Inform. 23 (3) (2019) 1304–1315.

[8] R. Patni, A challenge in cervical cancer elimination goal! J. Mid-life Health 13 (3) (2022) 263–264.

[9] L. Fasihi, B. Tartibian, R. Eslami, Artificial intelligence used to diagnose osteoporosis from risk factors in clinical data and proposing sports protocols, J. Orthop. Surg. Res. (2022).

[10] S. Lee, E.K. Choe, H.Y. Kang, J.W. Yoon, H.S. Kim, The exploration of feature extraction and machine learning for predicting bone density from simple spine X-ray images in a Korean population, J. Int. Skeletal Soc. A J. Radiol. Pathol. Orthop. 49 (2020) 613–618.

[11] M.A. Clynes, et al., The epidemiology of osteoporosis, Br. Med. Bull. 133 (2020) 105–117.

[12] C.L. Gregson, D.J. Armstrong, J. Bowden, C. Cooper, J. Edwards, N.J.L. Gittoes, UK clinical guideline for the prevention and treatment of osteoporosis, Arch. Osteoporos. 17 (58) (2022).

[13] G. Adani, T. Filippini, L.A. Wise, T.I. Halldorsson, L. Blaha, M. Vinceti, Dietary intake of acrylamide and risk of breast, endometrial, and ovarian cancers: a systematic review and dose-response meta-analysis, Cancer Epidemiol. Biomark. Prev. 29 (6) (2020) 1095–1106.

[14] K.K. Brieger, et al., Menopausal hormone therapy prior to the diagnosis of ovarian cancer is associated with improved survival, Gynecol. Oncol. 158 (3) (2020) 702–709.

[15] A.W. Lee, et al., Estrogen plus progestin hormone therapy and ovarian cancer: a complicated relationship explored, Epidemiology 31 (3) (2020) 402–408.

[16] T. Liu, Y. Song, R. Chen, R. Zheng, S. Wang, L. Li, Solid fuel use for heating and risks of breast and cervical cancer mortality in China, Environ. Res. 186 (2020).

[17] Y. Suzuki, Y. Huang, J. Ferris, A. Kulkarni, D. Hershman, J.D. Wright, Prescription of hormone replacement therapy among cervical cancer patients with treatment-induced premature menopause, Int. J. Gynecol. Cancer 33 (2023) 26–34.

[18] A. Brennan, M. Rees, Menopausal hormone therapy in women with benign gynaecological conditions and cancer, Best Pract. Res. Clin. Endocrinol. Metab. 35 (6) (2021).

[19] K. Agarwal, K.E. Cherian, N. Kapoor, T.V. Paul, OSTA as a screening tool to predict osteoporosis in Indian postmenopausal women—a nationwide study, Arch. Osteoporosis 17 (2022).

[20] E.S. Siris, et al., The effect of age and bone mineral density on the absolute, excess, and relative risk of fracture in postmenopausal women aged 50–99: results from the National Osteoporosis Risk Assessment (NORA), Osteoporos. Int. 17 (2006) 565–574.

[21] S.M. Cadarette, et al., Development and validation of the osteoporosis risk assessment instrument to facilitate selection of women for bone densitometry, Can. Med. Assoc. J. 162 (9) (2000) 1289–1294.

[22] T.K. Yoo, S.K. Kim, D.W. Kim, J.Y. Choi, Osteoporosis risk prediction for bone mineral density assessment of PostMenopausal women using machine learning, Yonsei Med. J. 54 (6) (2013) 1321–1330.

[23] A.C. Antoniou, P.P.D. Pharoah, P. Smith, D.F. Easton, The BOADICEA model of genetic susceptibility to breast and ovarian cancer, Br. J. Cancer 91 (2004) 1580–1590.

[24] A.C. Antoniou, et al., The BOADICEA model of genetic susceptibility to breast and ovarian cancers: updates and extensions, Br. J. Cancer 98 (2008) 1457–1466.

[25] A. Lee, et al., BOADICEA: a comprehensive breast cancer risk prediction model incorporating genetic and nongenetic risk factors, Genet. Med. 21 (2019) 1708–1718.

[26] M.X. Ji, Q. Yu, Primary osteoporosis in postmenopausal women, Chronic Dis. Transl. Med. 1 (1) (2015) 9–13.

[27] N. Kliemann, et al., Predicted basal metabolic rate and cancer risk in the European prospective investigation into cancer and nutrition, Int. J. Cancer 147 (3) (2019) 648–661.

[28] J.C.M. Ng, C.M. Schooling, Effect of basal metabolic rate on cancer: a mendelian randomization study, Front. Genet. 12 (2021).

[29] L.S. Toh, et al., A comparison of 6 osteoporosis risk assessment tools among postmenopausal women in Kuala Lumpur, Malaysia, Osteop. Sarcop. 5 (3) (2019) 87–93.

[30] J.-G. Shim, D.W. Kim, K.-H. Ryu, E.-A. Cho, J.-H. Ahn, J.-I. Kim, S.H. Lee, Application of machine learning approaches for osteoporosis risk prediction in postmenopausal women, Arch. Osteop. 15 (2020).

[31] U. Ferizi, S. Honig, G. Chang, Artificial intelligence, osteoporosis and fragility fractures, Curr. Opin. Rheumetol. 31 (4) (2019) 368–375.

[32] S. Rani, A. Sehgal, J. Kaur, D.K. Pandher, R.S. Punia, Osteopontin as a tumor marker in ovarian cancer, J. Mid-life Health 13 (3) (2022) 200–205.

[33] R. Patni, A challenge in cervical cancer elimination goal! J. MidLife Health 13 (3) (2022) 263–264.

[34] L. Zeng, L. Liu, D. Chen, H. Lu, Y. Xue, H. Bi, W. Yang, The innovative model based on artificial intelligence algorithms to predict recurrence risk of patients with postoperative breast cancer, Breast Cancer, Sect. J. Front. Oncol. 13 (2023).

[35] Z. Ma, Y. Yang, J. Lin, X.D. Zhang, Q. Meng, B.Q. Wang, BFH-OST, a new predictive screening tool for identifying osteoporosis in postmenopausal Han Chinese women, Clin. Interv. Aging 11 (2022) 1051–1059.

[36] U. Ferizi, S. Honig, G. Chang, Artificial intelligence, osteoporosis and fragility fractures, Curr. Opin. Rheumatol. 31 (4) (2020) 368–375.

[37] E. Lydick, K. Cook, J. Turpin, M. Melton, R. Stine, C. Byrnes, Development and validation of a simple questionnaire to facilitate identification of women likely to have low bone density, Am. J. Manag. Care 4 (1) (1998) 37–48.

[38] S. Fujiwara, et al., Performance of osteoporosis risk indices in a Japanese population, Curr. Therap. Res. Clin. Exp. 62 (8) (2001) 586–594.

[39] P. Geusens, et al., Performance of risk indices for identifying low bone density in postmenopausal women, Mayo Clin. Proc. 77 (7) (2002) 629–637.

[40] C. Suzzane, et al., Validation of various osteoporosis risk indices in elderly Chinese females in Singapore, Osteoporos. Int. 17 (8) (2006) 1182–1188.

[41] M. Cadarette, et al., Development and validation of the osteoporosis risk assessment instrument to facilitate selection of women for bone densitometry, CMAJ—Can. Med. Assoc. J. 162 (9) (2000) 1289–1294.

[42] Collaborative group on hormonal factors in Breast Cancer, Menarche, menopause, and breast cancer risk: individual participant meta-analysis, including 118 964 women with breast cancer from 117 epidemiological studies, Lancet Oncol. 13 (11) (2012) 1141–1151.

[43] C. Sharma, M. Sharma, R. Raina, S. Verma, Gynecological diseases in rural India: a critical appraisal of indications and route of surgery along with histopathology correlation of 922 women undergoing major gynecological surgery, J. Mid-Life Health 4 (2014) 55–61.

Chapter 10

Mammography screening of women in their forties: Benefits and risks

Jyotsana Suyal, Akash Samanta, Kiran Dobhal, and Vikash Jakhmola
Uttaranchal Institute of Pharmaceutical Sciences, Uttaranchal University, Dehradun, Uttarakhand, India

10.1 Introduction

10.1.1 About breast malignancy cancer

More women in India are affected by breast cancer than by any other type of cancer. In 1960 the 195 nations in the globe, breast cancer affects women the most frequently. In 2020, Belgium and the Netherlands had the highest rates of breast cancer in women [1,2]. Breast cancer is the second most frequent cause of tumor-associated death in women. According to studies, India is expected to report more than 170,000 cases of breast malignancy in women by 2020. Research indicates that the condition may impact 1 in every 28 women. Almost 685,000 people worldwide died in 2020 because of breast cancers, including 2.3 million women [2,3]. Breast cancer was the world's most common disease in the 5 years leading up to 2020, with 7.8 million women receiving a diagnosis [3]. A family's genetic mutations that have been passed from generation to generation are responsible for 5%–10% of breast cancer cases [1–3]. The BRCA1 and BRCA2 genes, both of which significantly elevate the risk of breast and ovarian cancer, are the two most well-known hereditary mutant genes that potentially increase the chance of breast cancer [3].

In the glandular tissue of the breast, breast malignancy expands in the inner epithelial tissue of the ducts (85%) or lobules (15%) [2,3]. Breast cancer first manifests in breast tissue. Breast cells mutate and amplify, producing a mass of tissue [4]. Like other cancers, breast cancers have the potential to multiply to the tissue surrounding the breast. Additionally, the cancer can travel to different areas of the body and develop new tumors. Unlike normal cells, cancer cells can grow outside of the area of the body where they first appeared [4,5]. Women older than 50 years are more likely to develop breast cancer. Although

Artificial Intelligence *and* Machine Learning
for Women's Health Issues
https://doi.org/10.1016/B978-0-443-21889-7.00009-9

rare, breast cancers can also occur in men. Male breast cancers affect approximately 2640 men in the United States annually, accounting for less than 1% of global instances. In stage 1 breast cancer, the tumor is <2 cm in length and is restricted to the breast as a mass. The lump can range from 2 to 5 cm in size in stage 2 breast cancer and may spread to nearby lymph nodes. Breast cancer that has reached stage 3 has progressed to at least internal mammary lymph nodes and may be of any size [6]. In stage 4, breast cancer has migrated to supplementary tissue structures [3,4,7]. The most conventional sign of breast malignancy is a painful lump or increase in the size of the breast [8,9]. Depending on the tumor's features, therapy often entails mastectomy (surgical removal of the breast). There are various types of breast malignancy, often categorized according to where it originates. Some of these types include the following:

- **Invasive (infiltrating) lobular carcinoma:** This malignancy begins in the lobules of the breast, where milk is generated, and subsequently spreads to neighboring breast tissue. It contributes to 8%–12% of breast malignancy cases [8].
- **Inflammatory breast cancer:** This cancer is rare, severe, and infectious looking. Breast skin dimpling, pitting, and swelling are common signs, which are caused by obstructive cancer cells in the lymphatic channels beneath the skin [9].
- **Invasive (infiltrating) ductal carcinoma:** This type of malignancy begins in the milk-producing glands. This is the most predominant malignancy and cause 80% of all manifestations [8,9].
- **Triple negative breast cancer (TNBC):** TNBC, which accounts for about 15% of all breast cancers, is among the most challenging types of breast cancer to treat [10]. There are three markers, i.e., estrogen, progesterone, and human epidermal growth factor receptor-2 associated with other forms of breast cancer that are not present in triple negative breast cancer [8,9].
- **Paget's disease of the breast:** Paget's disease of the breast is a rare complication of breast cancer, involving the nipple and areola. In this type of malignancy, the skin of the nipple and areola exhibits eczema-like changes [9].
- **Lobular carcinoma in situ:** In this condition, abnormal cells develop in the breast lobules or milk glands. The abnormal cells are not considered to be breast cancer and can be surgically removed.
- **Ductal carcinoma in situ:** Ductal malignant growth in situ is also known as stage 0 breast cancer. It is a preinvasive breast cancer, in which the cells that line the ducts have changed into cancer cells but they have not spread into nearby breast tissue.

The primary clinical questions addressed in this study are: (1) What is screening and mammography? (2) What benefits come with mammography screening, and how do they alter with patient risk and age? (3) What, if any, are the negative effects of mammography screening? (4) What is known about how to take into account unique traits when making recommendations for breast cancer screening? (5) How many people are helped to decide on mammography screening in a knowledgeable manner? [11].

1. **Screening Test:** Testing and examinations used to identify diseases in persons who are asymptomatic are referred to as screening. Breast malignancy screening tests are used for early detection before symptoms manifest. Early discovery refers to diagnosing and treating a disease before symptoms manifest. Some critical factors in diagnosis of breast cancer include tumor size and distribution [12].
2. **Breast Cancer Screening Guidelines:** The American Cancer Society (ACS) developed mammography guidelines for women at average risk of breast cancer. Women are considered to be at average risk if they have not had chest radiation before the age of 30 years and do not have a personal or family history of breast cancer or a genetic mutation that increases risk of breast cancer [12,13]. The ACS guidelines recommend the following:
   - Women aged 40–44 years can begin annual screening.
   - Women aged 45–54 years should get yearly mammograms.
   - Women aged 55 years and older can choose between continuing with annual exams and transitioning to a mammogram every other year. Women must continue to be screened as long as they are in good health and are estimated to live for at least another 10 years.
   - All women should be aware of what to expect and what the mammography can and cannot do before undergoing screening [12].

10.2 Mammography

Mammography is a medical imaging process that uses X-rays to view inside the breasts, allowing for early detection of disease. There are three types of mammography: digital mammography, computer-aided detection (CAD), and breast tomosynthesis [13,14].

10.2.1 Digital mammography

Digital mammography, also known as full-field digital mammography (FFDM), is a type of mammography that uses solid-state detectors instead of X-ray film [12]. From a patient's point of view, undergoing a digital mammogram is very similar to having a conventional film screen mammogram. In digital mammography, the patient is exposed to less radiation than in other techniques. The images generated by FFDM are uploaded to a computer and reviewed by a radiologist. This new technology produces clearer images at faster speeds than other methods, and these images can be enhanced, which makes interpretation easier and more accurate [13,15].

10.2.2 Computer-aided detection

Computer-aided detection (CAD) methods scan digitized mammographic pictures for abnormal areas of density or mass to detect the presence of cancer. These techniques are at the forefront of screening. They highlight any areas of abnormality so the radiologist can conduct further analysis [15].

10.2.3 Breast tomosynthesis

Breast tomosynthesis is also known as three-dimensional (3D) mammography and digital breast tomosynthesis (DBT). It is an advanced technique involving low-dose X-rays to capture multiple images of the breasts from different angles and computer reconstructions to create 3D images of the breasts. In this approach, 3D breast image is corresponding to computerized axial tomography scanner (CT) image, at which point the material is reconstructed in 3-D utilizing a order of thin "slices" (countenances) [15]. Despite being a little more that secondhand in standard mammography for few conscience tomosynthesis wholes, the dissemination measure is still inside the FDA-certified dependable limits for mammogram dissemination. Some procedures use dosages that are completely corresponding to those secondhand in established mammography. Breast tomosynthesis aids in the early detection and diagnosis of breast cancer [13,15].

Mammography can be either screening or diagnostic [12,14].

10.2.4 Screening mammography

Screening mammography is for women at average risk of breast cancer, that is, those who do not have a personal or family history of breast cancer or a genetic mutation that increases risk of breast cancer. By determining early-stage, undiagnosed breast cancers, screening mammography reduces rates of mortality. Early recognition also gives patients more options for treatment [14,15]. Randomized controlled trials (RCTs) show that mammography can decrease rates of breast cancer mortality by at least 20%. Women between the ages of 50 and 74 years are the most likely to benefit from screening mammography.

BreastScreen Aotearoa is New Zealand's free national breast screening program for women aged 45–69 years. A doctor's approval is not required for screening mammography through BreastScreen Aotearoa or BreastScreen Australia, which is another organization that offers free screening. Before undergoing mammography, patients may have some questions and concerns. We discuss these in the sections that follow [16,17].

Process before screening mammography

On the day of the mammography screening, women are asked to refrain from using deodorant, perfume, creams, powders, and so on. If a patient has breast implants, they should be prepared for the screening to take longer than it does for patients without implants. Patients are encouraged to wear comfortable clothes that are easy to put on and take off [14,18].

Process during screening mammography

Using a specific X-ray apparatus, mammography obtains images of the breasts, which are compressed between two plates. Compression increases image clarity and reduces the need for high dosages of radiation [17,19].

Effects of screening mammography

Due to the compression of the breasts during screening, some individuals may experience discomfort, redness, or swelling [12,14].

Duration of screening mammography

A screening mammography ideally takes about 10–15 min. If images are unclear, the screening may take extra time [20]. A radiologist usually reviews and assesses the images and determines whether additional images are needed [12,19].

Specialists for screening mammography

Radiographers, mammographers, or medical image technologists are the specialists that perform screening mammography. A radiologist is also part of the team, as they are responsible for evaluating and interpreting the screening images [19,21].

10.2.5   Diagnostic mammography

When screening mammography detects a clinical abnormality, like a breast lump or nipple discharge, the patient is advised to undergo diagnostic mammography. A diagnostic mammogram is used to evaluate unexpected findings on a screening mammogram and to investigate suspicious changes in the breast [22].

10.3   Benefits of screening mammography

Mammography screening has several benefits, including reductions in breast cancer mortality, years lost to the disease, and treatment-related morbidity [23].

10.3.1   Reduced mortality

The effect of breast cancer screening on mortality rates has been calculated utilizing many tests. More hostile screening consistently emerged in minority deaths, which unites all arrangements that are classified as protective pieces of advice. It is called "humanness decline" when most deaths prevented due to a particular screening procedure are avoided rather than using a different defense or not hiding at all. When screening age ranges are expanded and screening rates increase, the overall mortality rate declines. When compared to no screening, annual screening mammography reduces mortality in women aged 40–84 years by 40% [18,23]. Several randomized studies have been conducted to gather data on the effectiveness of mammography screening. According to some of these studies, women aged 50–69 years who underwent screening saw a statistically significant decrease in mortality of 23% [16]. Offering annual screening to women aged 40–49 years reduces mortality rate from 12% to 29%. The death rate for women over 50 is correspondingly higher than that of women 40–49 years old at primary protection. The number necessary to screen (NNS) is the number of people that need to be screened for a given duration to prevent one death or adverse event. Because the percentage of breast cancer in earlier women is lower, the NNS decreases with age. One review determined NNS to be 753 for women aged 40–49 years, 462 for those aged 50–59 years, and 355 for those aged 60–69 years [23–25].

10.3.2 Reduced years of life lost

Due to their longer life expectancies, younger women receive greater benefits from breast cancer screening, especially in terms of reduced years of life lost. Women the one who are determined with breast malignancy in their 40s, frequently drop 30% of those who are in higher age. If a woman is determined accompanying breast cancer at age 42 and dies at age 48 a suggestion of correction 74 and dies at 80, she has wasted more be present at her existence and misplaced out on space to enhance and benefit from organization [16]. Although breast cancer risk increases with age, the increase does not match the decline with advancing age. In order to gain individual period of growth, 20 women in their forties should receive annual screening, when in fact 45 women in their 70s must endure semiannual screening. The financial impact of the supplementary existence age acquired from avoiding breast cancer fatality in more immature women as opposite to earlier women has not still happened fully intentional. It is fair to want that more immature active-age women accompanying kids will be more damaged by breast cancer death than earlier women that are more inclined be elderly from the labor [24].

10.3.3 Reduced treatment morbidity

Tumors found upon breast cancer screening are frequently benign, small, and do not include lymph nodes. This in turn affects prognosis with 5 year survival rate for local disease at 99%, regional disease at 86% (e.g., feathered lymph growth), and distant metastatic disease at just 27%. Treatment options are determined by cancer stage; higher-stage cancers require more intensive treatment. When women between the ages of 40 and 49 are not constrained, they are 2.5 times more likely to require chemotherapy, 4.6 times more likely to have their "feathery" knots cut, and 3.4 times more likely to undergo a mastectomy [18,23,26].

10.4 Risk of screening mammography

Overdiagnosis, false-positive results, anxiety and distress, discomfort during the procedure, and radiation damage are the most often mentioned concerns of mammography (Fig. 10.1) [23].

10.4.1 Overdiagnosis and overtreatment

Several studies have shown a connection between mammography screening and overdiagnosis. Overdiagnosis refers to finding cancer during screening that will never cause any symptoms [16]. These cancers may just stop growing or go away on their own. Obligatory and nonobligatory overdiagnosis can be distinguished from one another. Obligatory overdiagnosis becomes necessary to over diagnose a woman's illness when she passes away for another reason before her screen-detected cancer shows up clinically [27]. Nonobligated overdiagnosis occurs when a cancer that has been detected on a screen does not become clinically evident. Overdiagnosis is the identification of cancer that is asymptomatic. Treatments for an overdiagnosis cancer are referred to as overtreatment.

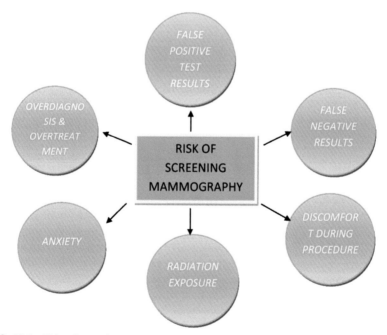

FIG. 10.1   Risks of screening mammography.

Although it is challenging to determine rates of overdiagnosis and overtreatment, some studies have been conducted [28].

Some women receive unnecessary therapy (overtreatment) due to overdiagnosis, even though the cancer would not have caused any problems. Occasionally, medical professionals are unable to discern between fatal and nonfatal cancers. Thus, they advocate for treating all breast cancers. As a result, some women receive cancer therapy when it is not actually necessary, and thus, they suffer the negative effects of treatment [12]. Although it is hard to calculate numbers of overdiagnoses of breast cancer, some studies suggest a rate of 10% in women older than 40 years. A 40-year-old woman's need for an unnecessary breast cancer diagnosis is anticipated to be 0.1% due to her high social expectations and low overall mortality.

Almost all women accompanied with breast cancer undergo next situation with surgery, radiation, birth control method therapy, or chemotherapy, in spite of the lack of correct assessments of overtreatment. Therefore, it is likely that overtreatment rates are identical to overdiagnosis estimates [18]. Unnecessary surgery, chemotherapy, and radiation healing are side property of overtreatment. There is no evidence that the harm caused by overtreating cancer disagrees from the harm caused by essential analysis, other than the event that the departed is not outweighed by a profit. For DCIS, chemotherapy is not secondhand [18,23].

10.4.2 False-positive test results

In the identification of breast cancer, there are two types of "false-positives": false-positive recalls and false-positive biopsies. A false-positive recall occurs when a woman is asked to come in for additional screening to further examine a problematic area that is subsequently found to be noncancerous. This happens in about 6.5% of cases of screening mammography [18,28]. In North America, 7.5% of women who have had screening are asked to come back for further assessment due to radiological symptoms, and following additional noninvasive imaging, they are informed that the results are benign or normal. This is usually described as a "false-positive" (FP) test without biopsy, despite the deceptive name. In around 1% of cases, women who have had a breast biopsy are asked to come back for additional screening even though their biopsy results are negative. Women who are younger, have dense breasts, have had previous breast biopsies, have a family history of breast cancer, and take estrogen are more likely to get false-positive results. Over a 10-year period, a few women who have yearly mammograms will experience a false-positive result. When it is the first mammogram, the chance of a false-positive result is greater (or unless there are previous mammograms to compare it against). Women who have access to their earlier mammography data can cut their risk of receiving a false-positive result in half. Mammography false-positives can cause anxiety. Ruling out malignancy, however, might require extra testing, which takes more time, money, and occasionally even suffering [16,18]. The recall rates for false-positive results are impacted by the age at which women start screening. Compared to women who start annual screening at age 40 years, who normally have one false-positive every 10 years, women who start at age 50 years typically experience one false-positive every 11.5 years. The recall rate is about 50% better on a guideline mammography, which is significant even when the strength of any potentially positive results cannot be determined. Hence, depending on the age a woman begins screening, there will be an increase in false-positive recalls before and after normalization [23,29,30].

10.4.3 False-negative results

Even when breast cancer is present, a mammography may produce a false-negative result. Screening mammographies typically miss one in eight cases of breast cancer. Women with big breasts have a greater risk of experiencing false-negative results. Women who have breast cancer may experience false-negative mammography results. Even if a patient has just had a normal screening mammography, it is crucial they visit their doctor if they experience any new breast problems. To examine the region more thoroughly, further tests, like a diagnostic mammography and/or a breast ultrasound, may be required [23,30].

10.4.4 Anxiety

Anxiety is generally reported following a screening recall due to possibility of breast cancer. Anxiety is also cited as harm of breast cancer screening because it is considered unnecessary when most women undergo screening are healthy. Women with completely false test results were less likely to undergo their subsequent screening mammography, according to two studies. Financial costs associated with false-positive test results must frequently be covered entirely or in part by the patient [21,28]. Some women mention that they are more aware of the threat of getting breast cancer, even after receiving false-positive results, and thus, they undergo screening more often because it makes them feel more comfortable and secure. Studies have shown that the fear of a false-positive mammogram was accompanied by a redoubled focus on the advantages of screening, imaging technology, and risk reduction. Courses that educate people about breast cancer screening and provide emotional support can help lessen the anxiety associated with breast cancer screening [16,19].

10.4.5 Discomfort during procedures

According to a study by the US Preventative Services Task Force, some women experience embarrassment during mammography, but this rarely causes them to avoid screening. The Task Force's review did not address diagnostic methods for fake-positive mammography findings, though they may also be embarrassing and awkward [28].

10.4.6 Radiation exposure

Radiation exposure is one potential drawback of screening mammography. A standard screening mammogram involves two views of each breast, and allied law sets a supervisory limit of 3.0 mGy for each mammography image of a breast of normal size. Current digital mammography and digital breast tomosynthesis radiation doses are frequently much lower than the permitted maximum. A woman went for annual mammography screening from the ages of 40–49 is concluded to have fateful fallout-persuaded breast malignancy approximately earlier per 76,000–97,000 years. Radiation care is crucial in radiology, but mammography screening should not be avoided due to the slight risk of radiation-caused injury [17,23].

10.5 Supporting informed decision-making

Discussions on the risks, benefits, doubts, options, and patient options concede possibility belong mammography accountable [31]. Few studies have examined indicators of an educated decision on whether to undergo screening, despite there being programs to promote and encourage mammography, including programs geared to a person's psychological preparedness to accept screening or to an individual's specific risk profile [11].

10.6 Conclusion

Breast cancer is a major health concern throughout most of Asia. However, rates of screening on the continent are not as high as they should be despite the availability of screening programs. Due to underutilization of existing public breast screening programs in developed countries, the full benefits of screening are not being realized. Women aged 40–49 years are more likely to experience fake-positive recall, fake-positive biopsies, and short-term anxiety than women aged 50 years and older. The primary benefits of screening are a decrease in mortality and morbidity and an increase in life expectancy. Younger women experience greater false-positive recalls and biopsies compared to older women, which is concerning. Radiation-induced cancer and overdiagnosis are two concerns that are rather uncommon at this time. Recalls and biopsies that trigger momentary fear. There are a lot of concerns in this age group regarding radiation-induced cancer and overdiagnosis which is highlighted in this chapter.

References

[1] S. Adams, M.E. Gatti-Mays, K. Kalinsky, Current landscape of immunotherapy in breast cancer, JAMA Oncol. 5 (8) (2019) 1205–1214.

[2] C. Anders, L.A. Carey, Understanding and treating triple-negative breast cancer, Oncology (Williston Park) 22 (11) (2008) 1233–1239.

[3] W.D. Foulkes, I.E. Smith, J.S. Reis-Filho, Triple-negative breast cancer, N. Engl. J. Med. 363 (20) (2010) 1938–1948.

[4] P. Kumar, R. Aggarwal, An overview of triple-negative breast cancer, Arch. Gynecol. Obstet. 293 (2016) 247–269.

[5] M. Shreshtha, A.B. Sarangadhara, S.D. Uma, S. Sunita, Epidemiology of breast cancer in Indian women, Asia Pac. J. Clin. Oncol. 13 (4) (2017) 289–295.

[6] D.A. Berry, Benefits and risks of screening mammography for women in their forties: a statistical appraisal, J. Natl. Cancer Inst. 90 (19) (1998) 1431–1439.

[7] J. Ferlay, M. Ervik, F. Lam, M. Colombet, L. Mery, M. Piñeros, A. Znaor, I. Soerjomataram, F. Bray, Global Cancer Observatory: Cancer Today, International Agency for Research on Cancer, Lyon, France, 2018. 3(20).

[8] B. Weigelt, F.C. Geyer, J.S. Reis-Filho, Histological types of breast cancer: how special are they? Mol. Oncol. 4 (3) (2010) 192–208.

[9] G.N. Sharma, R. Dave, J. Sanadya, P. Sharma, K. Sharma, Various types and management of breast cancer: an overview, J. Adv. Pharm. Technol. Res. 1 (2) (2010) 109.

[10] N.M. Almansour, Triple-negative breast cancer: a brief review about epidemiology, risk factors, signaling pathways, treatment and role of artificial intelligence, Front. Mol. Biosci. 9 (2022) 32.

[11] L.E. Pace, N.L. Keating, A systematic assessment of benefits and risks to guide breast cancer screening decisions, JAMA 311 (13) (2014) 1327–1335.

[12] C.D. Runowicz, C.R. Leach, N.L. Henry, K.S. Henry, H.T. Mackey, R.L. Cowens-Alvarado, R.S. Cannady, M.L. Pratt-Chapman, S.B. Edge, L.A. Jacobs, A. Hurria, American Cancer Society/American Society of Clinical Oncology breast cancer survivorship care guideline, CA Cancer J. Clin. 66 (1) (2016) 43–73.

[13] P.C. Gotzsche, K.J. Jørgensen, Screening for breast cancer with mammography, Cochrane Database Syst. Rev. 6 (2013).

[14] Inside Radiology, Available from: https://www.insideradiology.com.au/screening-mammography/.

[15] B.L. Niell, P.E. Freer, R.J. Weinfurtner, E.K. Arleo, J.S. Drukteinis, Screening for breast cancer, Radiol. Clin. North Am. 55 (6) (2017) 1145–1162.

[16] M.J. Yaffe, R.A. Jong, K.I. Pritchard, Breast cancer screening: beyond mortality, J. Breast Imaging 1 (3) (2019) 161–165.

[17] K.M. Ray, B.N. Joe, R.I. Freimanis, E.A. Sickles, R.E. Hendrick, Screening mammography in women 40–49 years old: current evidence, Am. J. Roentgenol. 210 (2) (2018) 264–270.

[18] A. Qaseem, J.S. Lin, R.A. Mustafa, C.A. Horwitch, T.J. Wilt, Clinical Guidelines Committee of the American College of Physicians*, Screening for breast cancer in average-risk women: a guidance statement from the American College of Physicians, Ann. Intern. Med. 170 (8) (2019) 547–560.

[19] D.B. Kopans, Misinformation and facts about breast cancer screening, Curr. Oncol. 29 (8) (2022) 5644–5654.

[20] S. Kour, R. Kumar, M. Gupta, Study on detection of breast cancer using Machine Learning, in: 2021 International Conference in Advances in Power, Signal, and Information Technology (APSIT), IEEE, 2021, pp. 1–9.

[21] K. Kerlikowske, J. Barclay, Outcomes of modern screening mammography, J. Natl. Cancer Inst. Monogr. (22) (1997) 105–111.

[22] V.P. Jackson, Diagnostic mammography, Radiol. Clin. N. Am. 42 (5) (2004) 853–870.

[23] L.J. Grimm, C.S. Avery, E. Hendrick, J.A. Baker, Benefits and risks of mammography screening in women ages 40 to 49 years, J. Prim. Care Community Health 13 (2022) 215–220.

[24] O. Mandrik, N. Zielonke, F. Meheus, J.L. Severens, N. Guha, R. Herrero Acosta, R. Murillo, Systematic reviews as a 'lens of evidence': determinants of benefits and harms of breast cancer screening, Int. J. Cancer 145 (4) (2019) 994–1006.

[25] P.C. Gøtzsche, O. Olsen, Is screening for breast cancer with mammography justifiable? Lancet 355 (9198) (2000) 129–134.

[26] M.M. Home, Screening mammography: do the benefits always outweigh the harms? Clin. Adv. Hematol. Oncol. 12 (6) (2014) 407–413.

[27] A. Maschke, M.K. Paasche-Orlow, N.R. Kressin, M.A. Schonberg, T.A. Battaglia, C.M. Gunn, Discussions of potential mammography benefits and harms among patients with limited health literacy and providers: "Oh, there are harms?", J. Health Commun. 25 (12) (2020) 951–961.

[28] American College of Obstetricians and Gynecologists, Breast cancer risk assessment and screening in average-risk women, Practice Bulletin (179) (2017) 245–252.

[29] C. Canelo-Aybar, M. Posso, N. Montero, I. Solà, Z. Saz-Parkinson, S.W. Duffy, M. Follmann, A. Gräwingholt, P. Giorgi Rossi, P. Alonso-Coello, Benefits and harms of annual, biennial, or triennial breast cancer mammography screening for women at average risk of breast cancer: a systematic review for the European Commission Initiative on Breast Cancer (ECIBC), Br. J. Cancer 126 (4) (2022) 673–688.

[30] Y.X. Lim, Z.L. Lim, P.J. Ho, J. Li, Breast cancer in Asia: incidence, mortality, early detection, mammography programs, and risk-based screening initiatives, Cancers 14 (17) (2022) 4218.

[31] M. Løberg, M.L. Lousdal, M. Bretthauer, M. Kalager, Benefits and harms of mammography screening, Breast Cancer Res. 17 (1) (2015) 1–2.

Chapter 11

Machine learning approach for early prediction of postpartum depression

Srishti Morris and Dipika Rawat
Uttaranchal University, Dehradun, India

11.1 Introduction

Postpartum depression (PPD) is a disorder that has serious negative impacts on both mothers and their children. Mothers may find it difficult to develop a close bond with their children, struggle to care for them, or even have bitter feelings toward them. The mother-infant attachment can be hampered by PPD, which could be detrimental to a child's growth. PPD is characterized as a period of moderate-to-severe depression during pregnancy or up to one year after delivery. It affects 8%–15% of women giving birth to their first child [1,2]. Both the mother and the child adjust to one another throughout the postpartum phase. In the initial 2 weeks after giving birth, feelings of anxiety, impatience, and sadness are common. These feelings usually pass on their own. Major and mild depression, which vary in intensity and prognosis, are frequently referred to as "postpartum depression."

In the early stage of pregnancy, postpartum depression can be controlled to some extent by using machine learning (ML) techniques. ML, a branch of artificial intelligence, describes techniques for predicting and plotting the trajectories of a range of unknown occurrences by identifying latent principles and distributions from input data [3].

11.1.1 Signs and symptoms of postpartum depression

See Fig. 11.1.

Artificial Intelligence *and* Machine Learning
for Women's Health Issues
https://doi.org/10.1016/B978-0-443-21889-7.00007-5

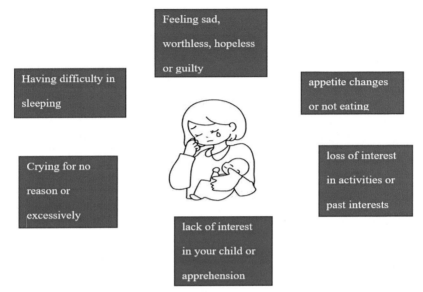

FIG. 11.1   Signs and symptoms of postpartum depression.

11.1.2   Causes of postpartum depression

Although the exact cause of PPD is unknown, the following factors may contribute [4]:

- hormonal changes
- physical changes
- history of depression in family
- stress (for caring baby)

As per neuroendocrinology studies, PPD causes erratic changes in the brain and women with PPD have intense changes in hypothalamic-pituitary-adrenal axis activities. An increase and decrease of hormones also causes PPD. Estrogen, progesterone, thyroid hormone, testosterone, corticotropin-releasing hormone, endorphins, and cortisol are among these hormones. After birth, estrogen and progesterone level drops within 24 h, also contributing to PPD.

In males, PPD can occur due to lifestyle changes and disrupted sleeping habits after the birth of the child (Fig. 11.2).

11.1.3   Types of postpartum depression [5,6]

(1)   Postpartum blues

Postpartum blues or "baby blues" is a typical condition following child-birth. Between 50% and 75% of new mothers experience postpartum blues after giving birth. The condition typically manifests itself in the first week.

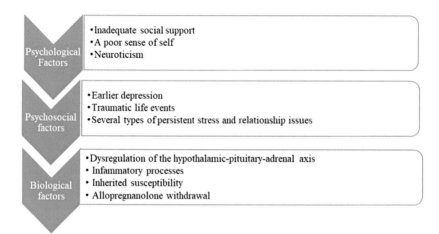

FIG. 11.2 The causes of postpartum depression.

Causes of postpartum blues:

The causes are not clearly established, however, the condition may occur due to biological and physiological changes (fatigue after childbirth, improper sleep, extra attention to the newborn, and anger and guilt due to preterm birth) during childbirth or due to family history.

Symptoms:

The common symptoms of postpartum blues are:

- mood swings
- crying often over long stretches of time and for no apparent cause
- anxiousness
- unhappiness

Treatment:

PPD typically goes away on its own in two weeks without therapy. However, the following may help mitigate the condition:

- proper sleep
- proper time to relax and enjoy
- family support
- avoiding alcohol and other drugs

(2) **Postpartum psychosis**

Postpartum psychosis is also called puerperal psychosis or peripartum psychosis. Short psychotic disorder is the term used to describe this disease. Postpartum psychosis is an extremely serious issue compared to PPD that requires rapid medical attention. It is extremely rare, affecting only 1 in 1000 people after birth. The signs and symptoms frequently appear immediately after birth, are acute, and exist for several days to months. Postpartum psychosis requires prompt medical attention due to the increased risk of suicide and potential injury to the child.

Causes of postpartum psychosis:

The most contributing factors to the onset of postpartum psychosis are hormonal and immune changes. It commonly occurs in persons with bipolar disorder or schizophrenia. It can also be a genetic disorder or caused by circadian rhythm disruption.

Symptoms:

The most common symptoms of postpartum psychosis are:

- severe agitation
- disorientation
- delusions or hallucinations
- sleeplessness
- paranoia
- hyperactivity
- fast speaking

Treatment:

Medication, counseling, and hospitalization are frequently used as treatments. Improvement in sleep patterns also plays an important role in treatment. Electroconvulsive therapy and cognitive behavioral therapies are also used to treat postpartum psychosis. Medicines like benzodiazepines, lithium, and other antipsychotic drugs are also used in treatment.

(3) **Postpartum depression**

Postnatal depression and PPD are both terms for the same condition. PPD is related to childbirth and may affect both parents. It affects roughly 15% of females and 1%–26% of males. It is a more dangerous disorder than the "baby blues." If the parents have a history of depression, then the chance of developing PPD increases to 30%. PPD is a depressive disorder with peripartum onset. Peripartum onset starts any time during the period of pregnancy or after delivery, within 4 weeks of birth or even after 1 year.

Symptoms:

The symptoms of PPD are:

- sadness
- mood swings
- frustration and helplessness
- guilt and shame
- thoughts of self-harm
- lack of interest
- low libido
- fatigue
- poor self-care
- harming partner or baby

Treatment:

- Psychotherapy or antidepressants are extremely effective forms of treatment. Emotional support by partner and family also helps in the prevention of PPD.

- Cognitive behavioral therapy, interpersonal therapy, and hormonal therapy are also effective in the prevention of PPD.
- Medicines like selective serotonin reuptake inhibitors (SSRIs) are effective for treatment. Brexanolone, zuranolone, and ganaxolone are prescription drugs used to treat PPD.

11.2 Assessment of postpartum depression by machine learning

ML is a technique used to predict clinical disorders via interpretation of large amounts of data. ML assessment can be accomplished by the following methods:

- **Supervised learning:** This refers to the process of training models to forecast the labels of fresh data using labeled data.
- **Unsupervised learning:** This is a technique for uncovering dependable internal patterns from unlabeled data.
- **Reinforcement learning:** This includes making judgments based on the laws of a certain environment by using learning algorithms [7].

Mild PPD episodes might be difficult to identify. Thus, replies to questions posed by healthcare professionals are very important. Medical professionals routinely utilize the Edinburgh Postnatal Depression Scale (EPDS) to screen for PPD. This scale includes 10 questions that are related to depressive symptoms, including feeling down, worried, or guilty. The response that most closely matches how the person felt during the past week should be checked. A higher score suggests that PPD may exist.

ML is being used to enhance the precision of PPD screening and early detection of women at risk using data from primary care electronic health records (EHRs). An artificial intelligence (AI) model that was developed utilizing a range of sociodemographic and medical variables was used to predict the likelihood of PPD in the first year of birth. The EPDS was evaluated as a standalone tool for accuracy and as an addition to the typical questionnaire-dependent screening to evaluate the model's performance using a variety of validation techniques.

A quantitative and objective tool for PPD screening is available in the form of PPD risk prediction using EHR data. This tool helps to identify PPD prior to pregnancy more precisely and aids in the identification of women at risk [8].

11.3 Machine learning-based postpartum depression prediction studies

ML algorithms have successfully predicted the persistence, chronicity, severity, and treatment response of major depressive illness. Studies have discussed the use of EHRs, mood rating scales, brain imaging data, smartphone monitoring systems, and social media platforms to predict, classify, or subgroup mental health conditions like depression, schizophrenia, and suicidal ideation and attempt [2].

Several modules for identifying PPD have been prepared using ML. These algorithms can accurately identify PPD in the first week after birth. By overcoming the limitations of traditional statistical views, ML techniques may be a viable way to correctly categorize and analyze critical complex data from the digital information received during pregnancy and in the postpartum period. Particularly in clinical practice, implementing screening techniques based on EHRs may simplify and improve the information of patients who are at risk.

The use of massive data to create a single model and assess the model's predictive power for persons who had not previously been seen, such as at-risk and new patients, has made ML-based predictive models increasingly popular to create systems that can automatically learn from data. ML approaches employ sophisticated statistical and probabilistic techniques. As a result, it is now feasible to more quickly and precisely spot patterns in data and forecast the future using data sources (e.g., more accurate diagnosis and prognosis).

Different ML algorithms model data in various ways. We compare the following algorithms:

1. **Support vector machine (SVM):** Through this technique, classification and regression issues can be addressed. It is a supervised ML technique. It is categorized as data points to identify a hyperplane in an N-dimensional space and separate the two groups of data points. It depends on the number of characteristics. For instance, the hyperplane is only a line if there are just two features. Finding the plane with the biggest margin, or the utmost distance between data points for both classes, is the objective.

   The SVM method, which minimizes structural risk in a collection of currently accessible data, is a supervised learning technique used to provide PPD predictions. It distinguishes between two classes that use the supplied data to produce a multidimensional hyperplane.

2. **Distributed Radom Forests (DRFs):** This is an algorithm for supervised ML that is used to solve classification and regression issues. To locate a hyperplane in an N-dimensional space and divide the two classes of data points, it is classified as data points. It depends on how many features there are. For example, if there are just two features, the hyperplane is just a line. The objective is to locate the plane with the highest margin or the utmost separation between data points for both classes. A supervised learning technique used to create PPD predictions is the SVM algorithm, which minimizes structural risk within a set of data that is currently accessible. It distinguishes between two types that create multidimensional hyperplanes from input data.

3. **Extremely Randomized Trees (XRTs):** The splitting criterion works by constructing multiple trees, which results in a lower variance but a greater

bias. XRTs are equivalent to DRFs, with the exception that there is higher volatility. A meta-learner is trained to determine the ideal configuration of base learners using a set of data and techniques and a subset of supervised learning techniques. Unlike bagging and boosting, where the idea is to group a number of weak learners together, the purpose is to combine a range of strong learners together to maximize learning [9].

4. **Naïve Bayes (NB):** NB is a classifier worked on probabilities and assumption that, for a given outcome, the presence of one feature does not automatically infer the presence of another feature. To be clear, regardless of whether they depend on one another or on the accessibility of other features, the NB assumes that each component separately affects the chance of an occurrence. To train a new model, stacked ensemble combines the predictions of earlier models.

5. **Ridge Regression:** This is a technique of analyzing multiple regression data with multicollinearity.

6. **LASSO Regression:** This, like linear regression, lessens the data near to the central points (i.e., models with fewer parameters).

7. **Gradient Boosting Machine (GBM):** GBM specifies a group of weak, thin succeeding trees, each learning from and refining upon the preceding trees [10].

8. **Computerized clinical decision support system (CDSS):** A CDSS for Android mobile applications used the model that best matched specificity and sensitivity. This method enables early estimation and diagnosis of PPD since it fits the criteria for an efficient test with a decent degree of sensitivity and specificity, is quick to perform, easy to interpret, culturally sensitive, and cost-effective. Future studies could evaluate the mobile app clinically. Android (Google, Mountain View, CA) has become one of the current mobile platforms with the biggest market penetration for the development of these types of apps, activating more than 1 million devices per day. Moreover, it provides depositories to develop powerful tools that are used by engineers [2,11].

Case study 1

Sampling: The results of 1397 patients' responses to psychiatric questionnaires, clinical data, and socioeconomic characteristics were collected in a database. Both new mothers and doctors who wish to keep track of their patients' tests can use an m-health app that connects with the Android platform, the model with the greatest performance, to collect data. To create models to forecast PPD over the 32 weeks of delivery, biological and psychosocial factors were examined, including data from the 8th week after birth and a genomic DNA study.

Each participant was White, had never received psychiatric treatment while pregnant, and was able to read and complete the clinical questionnaires. Mothers who lost a child after giving birth were not included. All the patients provided written informed permission for this investigation, which was authorized by the principled research committees.

With ML and precision recovery (PR), characteristics are automatically found in the data and classified into several types or categories. PR modules are often built using ML methodologies, which provide statistical and computational techniques to conclude information from the data in specific domain.

The primary stages of a PR issues based on ML are training and recognition. The PR model is constructed using a collection of data during the training phase. To solve fresh samples during the recognition phase, an adaptive module is tuned at this point for the optimal generalization. After the module is completed, a CDSS may use to help subsequent observations identify these patterns. Age, education level, labor situation during pregnancy, and employment status were independent factors included in the study.

Evaluation: Four alternative algorithm classifiers have been used in this study to predict postpartum outcomes. These alternative algorithm classifiers are:

- naive Bayes
- logistic regression
- support vector machine (SVM)
- artificial neural networks (ANNs)

After information analysis and the practice of a data mining technique for the creation, validation, and evaluation of multiple organized models, the NB model performed well, with balanced sensitivity and specificity close to 0.73. Mothers can provide answers to queries regarding the independent variables employed at the classification models throughout the first week following birth.

The case study was combined with a clinical decision assistance software for mobile health on the Android platform. Both new mothers and doctors who wish to keep an eye on their patients' test results can use this smartphone app. The risk of PPD is identified if the user responds to all the test questions. An affordable, simple test with verifiable findings that can be done after the first week of delivery on most Android mobile devices developed with a respectable level of accuracy. This is related to the concept of an effective and affordable screening since it would enable the detection of potential PPD cases that would otherwise go undetected through the response to a few straightforward questions.

Several models have been created using ML methods to predict PPD. These models can reasonably predict PPD within the first week of childbirth. Ultimately, a CDSS Android mobile app was created using the model that best balanced sensitivity and specificity. The app is quick to run, simple to interpret, culturally sensitive, and cost-effective, and can enable the early prediction and diagnosis of PPD. Future research may involve a clinical evaluation of the smartphone app.

Case study 2

The health of mothers and children is affected by PPD; however, many women do not obtain the proper therapy. For high-risk women, preventive treatments are cost-effective, but we are not very good at identifying them. The clinical, demographic, and psychometric information required to determine if PPD can be accurately predicted by ML techniques. In Uppsala, Sweden, between 2009 and 2018, a cohort study (BASIC study, $n=4313$) on population-based prospective was conducted. Women with a history of depression underwent sub analyses in this investigation. Maximum precision, sensitivity, and specificity was achieved by the randomized trees technique (73% accuracy, 72% sensitivity, 75% specificity, 33% positive predictive value, 94% negative predictive value, and 81% area under the curve). Accuracy was 64% among women without prior mental health concerns [9].

11.4   Challenges

Less than 10% of the women in the BASIC dataset experienced PPD, making it a population-based sample that does not accurately reflect clinical circumstances. This extreme data class imbalance is because most of the data was collected from women not showing symptoms of PPD 6 weeks post childbirth. The results reported by ML experts trained on such uneven data are typically skewed. During ML training, the lower class of women with PPD were oversampled in order to lessen this imbalance. This strategy eliminates information loss as opposed to undersampling a higher class of females not having the issues of PPD [9].

Currently, the research on PPD prediction using ML approaches is limited, and the majority of studies use diverse variable selection and ML procedures, which reduces the general ability to understand the results or data.

11.5   Future work

Work is ongoing to develop advanced machines and techniques that can evaluate compound and big data to perform complex calculations to aid in the early diagnosis of PPD. More clinical research partnerships are necessary to develop accurate and effective ML prediction algorithms.

ML approaches could offer an appropriate strategy to accurately categorize and evaluate crucial complicated data, surpassing the constraints of traditional statistical viewpoints, given the current increase of digitalized and statistical information obtained at the time of birth and in the postpartum period. Implementing screening methods based on EHRs into clinical practice, in particular, might streamline and improve the identification of patients who are at risk of PPD.

Antidepressants that are generally used to treat PPD include:

- SSRIs like sertraline and fluoxetine
- Serotonin and norepinephrine reuptake inhibitors (SNRIs) like duloxetine and desvenlafaxine
- Bupropion
- Tricyclic antidepressants (TCAs) like amitriptyline and imipramine

References

[1] C.J. Pope, D. Mazmanian, Breastfeeding and postpartum depression: an overview and methodological recommendations for future research, Depress. Res. Treat. 2016 (2016).

[2] P. Cellini, A. Pigoni, G. Delvecchio, C. Moltrasio, P. Brambilla, Machine learning in the prediction of postpartum depression: a review, J. Affect. Disord. 309 (2022).

[3] K.S. Betts, S. Kisely, R. Alati, Predicting postpartum psychiatric admission using a machine learning approach, J. Psychiatr. Res. 130 (2020) 35–40.

[4] C.E. Schiller, S. Meltzer-Brody, D.R. Rubinow, The role of reproductive hormones in postpartum depression, CNS Spectr. 20 (1) (2015) 48–59.

[5] D.E. Stewart, S. Vigod, Postpartum depression, N. Engl. J. Med. 375 (22) (2016) 2177–2186.

[6] J. Hopkins, M. Marcus, S.B. Campbell, Postpartum depression: a critical review, Psychol. Bull. 95 (3) (1984) 498.

[7] M. Zhong, H. Zhang, C. Yu, J. Jiang, X. Duan, Application of machine learning in predicting the risk of postpartum depression: a systematic review, J. Affect. Disord. 318 (2022).

[8] G. Amit, I. Girshovitz, K. Marcus, Y. Zhang, J. Pathak, V. Bar, P. Akiva, Estimation of postpartum depression risk from electronic health records using machine learning, BMC Pregnancy Childbirth 21 (1) (2021) 1–10.

[9] S. Andersson, D.R. Bathula, S.I. Iliadis, M. Walter, A. Skalkidou, Predicting women with depressive symptoms postpartum with machine learning methods, Sci. Rep. 11 (1) (2021) 1–15.

[10] K.P. Hirst, C.Y. Moutier, Postpartum major depression, Am. Fam. Physician 82 (8) (2010) 926–933.

[11] S. Jiménez-Serrano, S. Tortajada, J.M. García-Gómez, A mobile health application to predict postpartum depression based on machine learning, Telemed. e-Health 21 (7) (2015) 567–574.

Chapter 12

Improving women's mental health through AI-powered interventions and diagnoses

Rahul Negi

Kamla Nehru College, University of Delhi, New Delhi, India

12.1 Introduction

Women's mental health indicates a state of wellbeing in which a woman can manage the normal stress of life, work in different sectors productively, and make a positive contribution to society [1,2]. It encompasses a range of emotional, psychological, and social factors that affect the general health and wellbeing of women, and includes aspects such as self-esteem, relationships, work–life balance, and overall life satisfaction. Women's mental health also encompasses the diagnosis, treatment, and management of mental health conditions, including anxiety, depression, stress, and other disorders [3]. It is a critical aspect of a woman's overall health and wellbeing and requires ongoing attention and support to maintain and improve [4].

Women's mental health is a complex and multifaceted aspect of their general health that has a big impact on their quality of life. Women are frequently affected by mental health issues like depression, anxiety, and other disorders, which can have a significant impact on their everyday lives, interpersonal connections, and general happiness. Women's mental health is also shaped by societal, cultural, and environmental aspects such as gender-based violence and discrimination, work–life balance, and the stigma associated with mental health problems [5].

Despite its importance, women's mental health is often neglected or overlooked, and many women do not receive the support they need to maintain and improve their mental wellbeing. Early detection and intervention are critical to improving women's mental health, as well as increasing awareness and reducing the stigma associated with mental health conditions. The development of

Artificial Intelligence *and* Machine Learning
for Women's Health Issues
https://doi.org/10.1016/B978-0-443-21889-7.00017-8
173

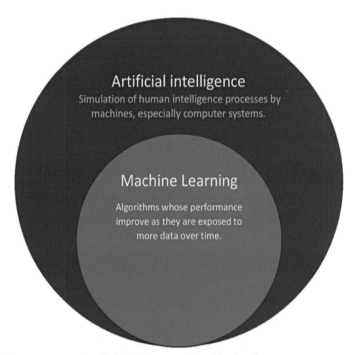

FIG. 12.1 Importance of artificial intelligence and machine learning.

innovative, AI-powered solutions for mental health diagnosis and intervention has the potential to make a significant impact on women's mental health and their wellbeing (Fig. 12.1).

Early detection and intervention of mental health concerns in women refers to identifying and addressing mental health issues in women as soon as possible after the onset of symptoms [6,7]. This can involve screening for mental health problems, early diagnosis, and prompt access to appropriate treatments such as therapy or medication. The goal of early detection and intervention is to improve outcomes and prevent mental health problems from becoming more severe, leading to a better quality of life for women and reduced stigma around mental health issues [8]. Early detection and intervention have the following beneficial effects:

- **Improved outcomes**: Early intervention can lead to better outcomes and a quicker recovery for women experiencing mental health problems.
- **Reduced severity:** Early detection can reduce the severity of mental health issues and prevent them from becoming more serious.
- **Increased access to treatment**: Early intervention provides women with greater access to appropriate treatments, such as therapy or medication, that can improve their mental health.

- **Decreased stigma**: By addressing mental health problems early on, women are more likely to feel comfortable seeking help, reducing the stigma surrounding mental health issues.
- **Improved quality of life**: Early intervention helps women manage their mental health, leading to improved quality of life and functioning in daily life.

The potential of AI to play an important role in improving women's mental health in the following areas:

- **Screening and diagnosis**: AI algorithms can assist in screening and diagnosing mental health issues in women, helping to identify problems earlier and speed up access to treatment.
- **Personalized treatment:** AI can analyze large amounts of data to develop personalized treatment plans for women based on their unique needs and characteristics.
- **Chatbots and virtual therapy**: AI chatbots and virtual therapy platforms can provide women with access to mental health support and resources, even if they live in areas without access to mental health professionals.
- **Mental health monitoring**: AI can be used to monitor women's mental health and provide real-time feedback, allowing for early detection and intervention of any issues that arise.
- **Research:** AI can be used to analyze large amounts of data related to women's mental health, helping researchers gain a better understanding of the causes and potential treatments for mental health issues in women.

AI has the potential to significantly increase the standard of treatment for women's mental health. To guarantee that AI-based solutions enhance and complement current human-led mental health services rather than replacing them, it is crucial to take into account ethical and privacy considerations when applying AI in mental health care.

12.2   AI-powered mental health diagnosis

The utilization of AI in mental health diagnosis is a topic of ongoing research and development. Some AI algorithms have been developed to assist mental health professionals in diagnosing certain mental health conditions. However, it is important to mention that AI should not replace the expertise and judgment of mental health professionals [9]. AI can provide additional information to support diagnosis, but a final diagnosis must be made by a trained mental health professional after considering all relevant factors, including the patient's history, symptoms, and behavior [10]. Additionally, AI systems are not yet advanced enough to replace the human aspect of mental health diagnosis, such as the ability to establish rapport and understanding with a patient, which is critical in the treatment of mental health.

The use of AI in mental health diagnosis is a growing field that aims to utilize AI techniques to improve the efficiency as well as accuracy of mental health assessments. AI algorithms can analyze large amounts of data, including medical history, patient symptoms, and other relevant information, to provide insights that may assist mental health professionals in making a diagnosis. Some AI models have been developed to diagnose specific mental health conditions, such as anxiety and depression, while others aim to provide more general assessments. It is important to mention that AI is not intended to replace mental health professionals but rather to complement and support their work [11]. Mental health diagnosis is a complex process that requires a thorough evaluation of a patient's circumstances and requires a human touch to build rapport and understanding with the patient. Currently, AI is still in the early stages of development for use in mental health diagnosis and more research is needed to determine its efficacy and limitations [12].

In conclusion, AI has the potential to positively impact mental health diagnosis, but it should be used in conjunction with the expertise of professionals and should not replace the human aspect of mental health treatment.

12.2.1 Advantages of AI in mental health diagnosis

Affordability—Users of AI-based and other mental health applications can obtain therapeutic help whenever and wherever they need it, in contrast to traditional counseling which requires scheduling and travel for visits. When compared to the costs of in-person counseling, lost work, the need to make other arrangements, and transportation, AI mental health services provide support at low or no cost [13].

Accessibility—AI-driven applications have the potential to diminish obstacles in accessing mental health care, such as shortages of personnel and limited availability of practitioners in remote and isolated areas. This becomes particularly important considering that more than 100 million individuals in the United States live in regions designated as Health Care Professional Shortage Zones. AI-powered chatbots and platforms with location-specific capabilities can provide continuous assistance whenever and for however long it is required.

Efficiency—AI algorithms for mental health care have already demonstrated efficacy in recognizing signs of depression, posttraumatic stress disorder (PTSD), and other disorders by analyzing behavioral signals. Additional studies have shown that algorithms are 100% accurate at identifying whether at-risk adolescents are likely to have psychosis and can detect anxiety-related behavioral signs with more than 90% accuracy. They also help patients in mental distress. For example, in a Woebot AI chatbot creators' randomized controlled experiment, consumers saw a marked decrease in hopelessness and anxiety after just 2 weeks of using the app.

Privacy and ease to open up—Because of the privacy and ease with which they may open up to AI-based therapists, people are less self-conscious when they can communicate unpleasant truths with them. This is especially important for those who might feel humiliated in face-to-face interactions because of stigma or a fear of being evaluated. The most forbidden topics include smoking, drinking, and sexual behavior, and more than a quarter of people lie to doctors about these things. Since a robot will not condemn them, many people find it easier to divulge the full extent of their behavior to one.

Help for therapists—AI's remarkable capacity to swiftly and effectively analyze vast amounts of data surpasses human capabilities. As a result, these algorithms play a crucial role in achieving more accurate diagnoses. Furthermore, they can identify early signs of distress by closely monitoring a patient's mood and behavior, enabling timely alerts to healthcare professionals, leading to swift adjustments in treatment plans. This can save the lives of suicidal people who require regular check-ins.

12.2.2   Limitations of AI in mental health diagnosis

The limitations of using AI in mental health diagnosis include:

- **Lack of human interaction:** AI cannot provide the personal interaction and empathy that is crucial in mental health evaluations. And due to a lack of human involvement, there emerge trust and safety issues in the patient's mind, which in turn leads to the rise of anxiety and hypertension.
- **Limited understanding of human emotions and behavior:** AI is not yet capable of fully understanding human emotions and behavior, which are important factors in mental health diagnosis. Human emotions are complex and heterogeneous. They are influenced by many factors, such as genetics, upbringing, culture, and personal experiences. This complexity makes it laborious for AI to accurately model and anticipate human emotions.
- **AI in data:** AI algorithms rely on big datasets to learn as well as make decisions. However, these datasets may contain biases that can result in AI systems producing biased outcomes, which can lead to biased results in diagnoses. For example, if an AI model is trained on data that only comprises men, it may not perform as well when applied to women.
- **Limited understanding of mental health:** AI is limited by our current understanding of mental health, and new research may change the way mental health is diagnosed and treated.
- **Need for validation:** AI results must be validated by a licensed mental health professional, and AI should not be used as the sole source of information for a diagnosis. AI systems that are not validated may be susceptible to errors or failures when faced with unexpected situations or changes in the environment.

- **Ethical concerns:** Incorporating AI in mental health care brings forth significant ethical considerations, particularly regarding data security and privacy. AI systems have the capacity to gather and analyze extensive personal data, leading to legitimate concerns about safeguarding privacy and protecting sensitive information. There is a risk that AI systems may be used unethically to monitor or control individuals without their knowledge or consent, so it is crucial to develop AI responsibly and ethically.

12.3 AI-powered mental health interventions

AI-powered mental health interventions aim to provide individuals with convenient and accessible mental health support and treatment using AI-powered technologies such as chatbots, virtual therapists, mindfulness and stress-reduction apps, mood-tracking systems, cognitive behavior therapy (CBT) programs, and predictive analytics [14,15]. Some examples include:

- **Chatbots**: AI-powered chatbots can provide patients with mental health support and information through messaging platforms. They can identify symptoms, offer coping mechanisms, and even connect patients with support systems and mental health helplines.
- **Virtual therapists**: Virtual therapy sessions with AI-powered virtual therapists can provide mental health support and treatment to individuals who may not have access to traditional therapy services.
- **Mindfulness and stress-reduction apps:** AI can be used to develop mindfulness and stress-reduction apps that use techniques such as meditation and breathing exercises to help individuals manage their mental health. Some of these apps include Insight Timer (for meditation), Waking Up (for mindfulness), Happify (for CBT), Talkspace (for psychiatry).
- **Cognitive behavior therapy:** AI can be used to deliver CBT in a virtual format, allowing patients to receive treatment from the comfort of their own homes. By segmenting overwhelming problems into manageable pieces, CBT aims to help patients cope with them more constructively. Patients are taught how to alter undesirable thoughts so they can feel better. Contrary to numerous other "talk" treatments, CBT concentrates on the patient's present problems rather than problems from the past.
- **Mood tracking**: AI can also be used to track mood changes and provide personalized suggestions for mental health management. A mood tracker can identify patterns in your mood and factors that affect your emotions. You may begin recording your moods to better understand your emotions, identify undesirable actions, or find healthier coping mechanisms. It is often recommended by therapists to start tracking moods so that they can better comprehend the mental health of the patient and refine the treatment plan.

12.3.1 The advantages of AI-powered mental health interventions

Some advantages of AI-powered mental health interventions include the following:

- **Increased access to mental health services**: AI-powered treatments can advance access to mental health care, particularly for people who reside in distant or disadvantaged locations or who have trouble obtaining traditional therapy services [16].
- **Personalization**: AI-powered interventions can provide personalized treatment and support, tailoring interventions to individual needs and preferences.
- **Convenience**: AI-powered interventions can be accessed from the comfort of one's own home, reducing barriers to treatment and increasing accessibility.
- **Continuous support**: AI-powered interventions can provide continuous and ongoing support, allowing individuals to receive treatment and support at any time.
- **Data-driven decision-making**: AI-powered interventions can use various sources, such as mood tracking and wearable devices to gather data to inform treatment decisions and provide personalized recommendations.
- **Cost-effectiveness**: AI-powered therapies have the potential to be more affordable than conventional therapy, increasing access to mental health care for a larger spectrum of people.

    While AI-powered mental health interventions offer several advantages, they should not replace traditional therapy and must be used in conjunction with other evidence-based treatments. Additionally, the development and use of AI in mental health must be done ethically and responsibly to ensure patient safety and protect patient privacy and data security.

12.4 Challenges of mental health in women

Women encounter distinctive mental health obstacles shaped by a variety of biological, social, and cultural elements [17]. As per a new survey led by the Substance Abuse and Mental Health Services Administration, 29 million American women, or roughly 23% of the female populace, had encountered a diagnosable, psychological issue in the previous year. Furthermore, those are just the notable models. Certain mental health issues are more common in women and can significantly affect their general wellbeing [18]. While men have higher rates of chemical imbalance, beginning-stage schizophrenia, and alcoholism, women have higher rates of mental health issues, including the following:

- **Depression**: As compared to men, women are more likely to experience depression and other mood disorders, particularly during times of hormonal change, such as puberty, pregnancy, and menopause.

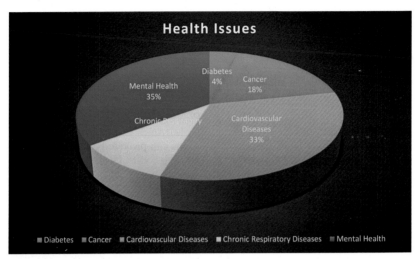

FIG. 12.2 Different health issues faced by women.

- **Anxiety disorders:** Generalized anxiety disorder, panic disorder, and phobias are among the anxiety disorders that affect women more frequently.
- **Eating disorders:** Females are at a greater risk than males for developing eating disorders like anorexia and bulimia, primarily due to societal pressures that impose unrealistic beauty standards upon them.
- **Trauma and abuse:** Women are more prone to suffer from trauma and abuse, such as sexual assault and domestic violence, which can have long-lasting effects on mental health.
- **Substance abuse:** Women may struggle with substance abuse and addiction, particularly in the context of trauma and abuse.
- **Reproductive health issues:** Women may experience mental health challenges related to reproductive health, such as postpartum depression, infertility, and menopausal symptoms.

It is worth mentioning that mental health challenges can affect anyone, regardless of gender, and seeking help from a mental health professional can make a significant difference in managing these challenges (Fig. 12.2).

12.5 Impact of societal and cultural factors on women's mental health

Societal and cultural factors can have a significant impact on women's mental health. Women's mental health is jeopardized by society's expectations of how they should behave, their duties in the family, gender prejudices, and so on [19,20]. Some of the psychological subjects that impact women's mental health include traumatic occurrences, abuse at an early age, an unhappy marriage, the stress of parenting, and family duties. The factors that deeply impact women's mental health include:

- **Gender roles and expectations**: Women are often expected to fulfill traditional roles, such as being the primary caretaker of children, managing the household, and providing emotional support to others, which can lead to high levels of stress and burnout.
- **Stereotyping and discrimination**: Negative stereotypes about women, such as the belief that women are weaker or less competent than men, can lead to discrimination and lower self-esteem, which can in turn contribute to mental health problems.
- **Sexual harassment and violence:** Women are frequently subjected to sexual harassment and abuse, which can result in PTSD, depression, and anxiety.
- **Body image issues**: Women are often subjected to unrealistic beauty standards, which can lead to body image issues and low self-esteem, which can contribute to mental health problems.
- **Lack of support**: Women may not have access to adequate support systems, such as flexible work arrangements, affordable childcare, or mental health services, which can make it difficult for them to manage their mental health.

Society needs to recognize the impact of these factors on women's mental health and work to address them. This can involve promoting gender equality, addressing negative stereotypes, and providing support and resources for women. Moreover, if women's general health is taken good care of, the upcoming generation will be healthy and may be unaffected by any negative impact of mental health conditions.

12.6   Significance of gender-specific methods in mental health

Gender-specific approaches to mental health therapies are essential because men and women encounter mental health challenges differently. In developing preventive strategies for mental health, it is crucial to consider both gender, which encompasses sociocultural constructs related to being male or female, and sex, which refers to biological characteristics. While gender and sex differences may be linked, gender encompasses a broader spectrum, interacting with other social factors, possibly amplifying biological susceptibilities unevenly. Considering gender and sex differences is vital due to their distinct and interconnected effects on mental health. This starts with implementing screening techniques to identify individuals at risk of mental health issues, preventing their onset, or intervening before they escalate. Some gender and sex differences in mental health include the following:

- **Different life experiences**: Women and men often face different life experiences, such as discrimination, harassment, and violence, which can impact their mental health differently.
- **Different coping mechanisms**: Women and men may use different coping mechanisms to deal with stress and mental health problems, and these differences may need to be taken into account in mental health interventions.
- **Different help-seeking behaviors**: Women and men may have different attitudes towards seeking help for mental health problems and may require different approaches to engage them in mental health services.
- **Different cultural and societal expectations**: Cultural and societal expectations can impact the way that men and women express their emotions and seek help for mental health problems.
- **Different biological factors**: There may be biological differences between men and women that impact their mental health, such as differences in hormone levels, brain chemistry, and susceptibility to certain mental health conditions.

Even though they may have the same mental health diagnosis, men and women may have different symptoms and hence require different types of therapy. For instance, a man who is sad is more likely to describe troubles with his employment, whereas a woman who is depressed is more likely to claim bodily problems, such as exhaustion or changes in eating or sleep patterns. Contrary to their male counterparts who also experience depression, women tend to experience alcohol abuse problems a few years following the onset of the condition. Women are more prone than males to seek release from their depressive symptoms through religious and emotional outlets, whilst men are more likely to seek comfort through sports and other interests.

By using gender-specific approaches in mental health interventions, health professionals can better understand and address the unique challenges faced by women and men in managing their mental health. This can lead to more effective and personalized treatment and ultimately better outcomes for patients.

12.7 Case studies of AI-powered mental health interventions for women

There have been several case studies on the use of AI-powered mental health interventions for women. Some examples include the following:

- **Woebot**: This is a chatbot-based program that uses CBT techniques to help users manage their mental health. Woebot is effective in reducing symptoms of anxiety as well as depression in women.
- **Ella**: This is a virtual mental health coach that uses AI to provide personalized help to women with depression, anxiety, and stress. Ella uses natural language processing (NLP) to understand users' needs and provide relevant support and resources.

- **Moodfit**: This is a mobile app that uses AI to deliver a personalized mental wellness program, including mindfulness and relaxation exercises, mood tracking, and goal setting. Moodfit has been used by women to manage stress and anxiety and improve mood and overall wellbeing.

These case studies show that AI-powered mental health interventions can be effective in helping women manage their mental health and improve their wellbeing. In any case, it is important to note that computer-based intelligence should not supplant professional mental health care, and looking for the guidance of a certified medical services provider is always recommended.

12.8   Mental health and physical activity during the COVID-19 pandemic

Proactive tasks and individual collaboration with partners, companions, and family have been limited due to COVID-19 measures like quarantine, lockdown, self-detachment, and social distancing [21]. During the pandemic, people experienced a notable decrease in overall wellbeing, a decline in mental health, and an increase in psychological distress, including stress, worry, sadness, and feelings of isolation. Such negative effects may have implications for the adherence to public health initiatives. To enhance mental health and wellbeing during this challenging time, it is imperative to conduct research and develop additional and alternative activities that can be pursued amidst the COVID-19 situation.

Engaging in exercise and physical activity is recognized for its ability to enhance or preserve both physical and mental health, ultimately improving overall quality of life. Several physiological and psychological factors contribute to the positive effects of physical activity. Exercise has been linked to increased levels of hormones, endorphins, and brain-derived neurotropic factor (BDNF), leading to a sense of happiness and reduced stress. Moreover, exercise provides individuals with an opportunity to take a break from stressors, and regular physical activity fosters a sense of self-efficacy in managing the challenges they encounter [22].

12.9   Discussion of the results and impact of the interventions

There are a variety of interventions that can be used to address mental health issues in women. Some examples include psychotherapy, medication, peer support groups, lifestyle changes, and complementary therapies. The results and impact of interventions will vary depending on the individual and the type of intervention used.

Psychotherapy: Psychotherapy, such as CBT, can be effective in treating a variety of mental health conditions in women, including depression and anxiety.

Research has shown that CBT can help women develop skills to manage negative thoughts and behaviors, leading to improved mental health outcomes.

Medication: Women with mental health disorders might benefit from medication, such as antidepressants and anxiety drugs. However, it is crucial to note that medicine may not be enough on its own and should be used with additional interventions like therapy.

Peer support groups: Peer support groups, such as those offered by organizations like the National Alliance on Mental Illness (NAMI), can provide women with social support and a sense of community, which can help manage mental health conditions.

Lifestyle changes: Working out, a good diet, and reducing stress are some life changes that can improve mental health. Research has demonstrated that engaging in physical activity can improve mood and mitigate symptoms of depression and anxiety.

Complementary therapies: Corresponding treatments, like needle therapy and yoga, can also help improve mental health. While research on these treatments is limited, some studies have shown that complementary therapies can reduce depression and anxiety.

The outcomes and effects of mediations on women's mental health depend on the type of medication used. Utilizing a mix of medications personalized to the patient will achieve the best results.

12.10 Potential future applications of AI in Women's mental health

AI has the potential to revolutionize the field of mental health of women in several ways. AI has enormous promise for improving mental health, particularly given the widespread usage of smartphones. The tools, if effectively planned and developed in partnership with medical professionals, can aid in early identification and evaluation, as well as recommend treatment alternatives. We discuss some potential future applications of AI in women's mental health in the sections that follow.

Personalized treatment plans: To create individualized treatment plans for women with mental health disorders, AI algorithms may be utilized to analyze information from a patient's medical history, genetic composition, and lifestyle variables.

Early detection and intervention: AI-powered screening tools can be used to detect mental health disorders in women earlier and provide timely interventions to prevent the conditions from worsening.

Virtual therapy sessions: AI-powered virtual assistants can provide women with mental health counseling and support, including mindfulness-based stress reduction, CBT, and other evidence-based interventions.

Natural language processing: AI algorithms can be used to analyze patient data from natural language sources, such as social media posts and text messages, to identify early warning signs of mental health issues and provide timely interventions.

Predictive analytics: AI can be used to predict which women may be at risk for developing mental health conditions, such as depression and anxiety, based on demographic, medical, and lifestyle data.

Wearable technology: Wearable gadgets, such as smartwatches and wellness trackers, can help monitor a women's health and provide appropriate measures. While smart technology has extraordinary promise for improving mental health, it must be developed in conjunction with mental health professionals.

12.11 Ethical considerations in using AI for women's mental health

Ethical considerations are crucial when using AI for women's mental health because they can help to ensure that the technology is developed and used in a way that is respectful, fair, and safe for all individuals. Women's mental health is a deeply personal and sensitive issue, and individuals have a right to privacy and confidentiality. Ethical considerations can help to ensure that AI is designed and implemented in a way that respects individuals' privacy and maintains the confidentiality of their data. There are several ethical considerations when using AI for women's mental health, including the following:

- **Data privacy**: AI-powered mental health interventions collect and store sensitive personal information. It is essential to make sure that this data is securely stored and shielded from unauthorized access.
- **Bias**: AI systems can perpetuate biases and discrimination if they are not trained on diverse and inclusive datasets. This can result in unfair or harmful outcomes for women, especially those from marginalized communities.
- **Accessibility**: AI-powered mental health interventions may not be accessible to all women, particularly those with limited access to technology or with disabilities.
- **Quality of care**: AI-powered mental health interventions may not provide the same quality of care as a human mental health professional. It is important to ensure that these interventions are backed by robust research and are appropriately regulated.
- **Responsibility**: The application of AI to mental health creates issues regarding accountability for any negative results. This includes issues such as confidentiality, informed consent, and accountability for any harm caused by the technology.

The ethical concerns underscore the significance of thoughtfully and cautiously employing AI in the realm of mental health, with a particular focus on women's wellbeing. Striking a balance between the potential benefits of AI and the imperative to protect the rights, privacy, and overall welfare of women becomes crucial. Careful attention and responsible implementation are essential to ensure that AI technologies in mental health are used ethically and equitably for the betterment of all individuals.

12.11.1   The importance of informed consent in AI-powered mental health interventions

Informed consent is crucial when using AI-powered mental health interventions, as it ensures that individuals understand the benefits and risks of the technology and make an informed decision about whether to participate. Key elements of informed consent include the following:

- **Explanation of the intervention**: Users should be given clear and concise information about the AI-powered mental health intervention, including its purpose, how it works, and its potential benefits and risks.
- **Data collection and use:** Users should be made aware of the types of data that will be gathered and their intended uses. Additionally, they should be made aware of who will have access to their data and how it will be safeguarded.
- **Potential risks and benefits**: Users should be informed about the potential benefits and risks of using AI-powered mental health interventions. This includes the potential for harm, such as privacy violations or inaccurate advice, as well as the potential for positive outcomes, such as improved mental health.
- **Alternatives**: Users should be informed about other available options, such as traditional mental health care, and should be allowed to make an informed decision about whether to participate in an AI-powered mental health intervention.
- **Right to withdraw**: Users should be informed that they have the right to withdraw from the AI-powered mental health intervention at any time and that this will not affect their access to other mental health services.

Informed consent is important for ensuring that individuals have control over their own mental health and wellbeing, and it is a critical component of ethical AI use in the field of mental health.

12.11.2   Ensuring privacy and data security in AI-powered mental health interventions

The privacy of a patient is of utmost importance in the healthcare industry. Patients need to trust their healthcare providers and feel safe in sharing personal information. When healthcare providers respect the patient's privacy, they build trust, which is critical to providing effective health care. Privacy ensures that a patient's personal information, including medical records, remains confidential.

Healthcare providers must maintain confidentiality to comply with legal requirements and to protect the patient's medical information from unauthorized access, use, and disclosure.

Ensuring privacy and data security in AI-powered mental health interventions requires implementing several measures such as the following:

- **Encryption of sensitive data**: All sensitive data should be encrypted during storage and transmission to prevent unauthorized admittance.
- **Access control**: Admittance to delicate information should be limited and controlled through secure validation strategies like passwords, biometrics, and multifaceted verification.
- **Data minimization**: The amount of sensitive data collected and stored should be minimized and only used for specific purposes related to the intervention.
- **Data retention policies**: The retention policies for sensitive data should be clearly defined and regularly reviewed to ensure that the data is not kept for longer than necessary.
- **Regular security audits**: Standard security reviews to distinguish and address any possible weaknesses in the framework should be conducted.
- **Compliance with relevant privacy regulations**: AI-powered mental health interventions should comply with relevant privacy regulations such as the General Data Protection Regulation GDPR in the European Union and the Health Insurance Portability and Accountability Act of 1996 (HIPAA) in the United States.
- **User awareness and education**: User awareness and education should be a priority to ensure that users understand the importance of privacy and security in AI-powered mental health interventions.

12.12 The potential impact of AI on women's mental health

Digital solutions, many of which include AI, offer promise for improving mental health. Tech businesses and universities are developing new tools with powerful diagnostic and treatment capabilities that may be utilized to serve vast populations at reasonable rates. AI and ML have fundamentally revolutionized disease detection, which has had a negative influence on women's mental health. The potential impacts of AI on women's mental health can be both positive and negative.

Positive impacts

- **Increased access to mental health resources**: AI-powered tools and interventions can increase access to mental health resources for women, especially those in underserved communities.
- **Improved diagnosis and treatment**: AI has the potential to enhance the precision of mental health diagnoses and offer personalized treatment recommendations.

- **Decreased stigma and discrimination**: AI-powered mental health interventions can reduce the stigma and discrimination associated with mental health issues.

Negative impacts

- **Privacy and security concerns**: The use of AI in mental health raises concerns about the privacy and security of sensitive data.
- **Bias in AI systems**: The data that AI systems are trained on determines how biased they are. If the data contains gender biases, AI systems may reinforce or amplify those biases in their recommendations.
- **Dependence on technology**: Overreliance on AI-powered mental health interventions may prevent women from developing the skills and self-awareness necessary for maintaining their mental wellbeing.

The effects of AI on women's mental health will be determined by the way these tools and interventions are created, applied, and utilized. It is crucial to thoroughly evaluate the potential advantages and risks, striving to develop and implement AI-powered mental health interventions in a manner that maximizes positive outcomes while minimizing any potential adverse effects.

12.12.1 Future directions for research and development in AI-powered mental health interventions

Future directions for research and development in AI-powered mental health interventions include the following:

- **Bias and fairness:** Addressing and reducing bias in AI systems to ensure that mental health interventions are accessible and effective for all individuals regardless of race, gender, and other demographic factors.
- **Integration with human-delivered care**: Integrating AI-powered interventions with human-delivered care to ensure that individuals receive the most effective and appropriate treatment for their needs.
- **Privacy and security**: Ensuring the privacy and security of sensitive data in AI-powered mental health interventions through the implementation of robust encryption and access control systems.
- **Personalized and adaptive interventions**: Developing AI-powered interventions that are personalized and adaptive, able to adjust to an individual's changing needs and preferences over time.
- **Evidence-based interventions:** Ensuring that AI-powered mental health interventions are evidence based and supported by rigorous scientific research.
- **Ethical considerations**: Addressing the ethical implications of AI-powered mental health interventions, such as informed consent, transparency, and accountability.

• **Interdisciplinary collaboration**: Encouraging interdisciplinary collaboration between experts in AI, mental health, and related fields to ensure the development of safe, effective, and accessible AI-powered mental health interventions.

12.12.2 The importance of continued efforts to improve women's mental health through AI and other means

AI and ML are playing a great role in finding means to solve various mental health issues faced by women worldwide. Continued efforts to improve women's mental health through AI and other means are important for several reasons.

• **Prevalence of mental health issues**: Given that women are more susceptible to specific mental health problems like anxiety, depression, and trauma-related disorders, it becomes imperative to develop effective and easily accessible mental health interventions tailored to their needs.
• **Improved quality of life**: Improving women's mental health can lead to higher productivity, lower healthcare expenditures, and better life quality.
• **Addressing disparities**: Mental health in women is often neglected and underfunded, and continued efforts are needed to address the disparities in access to mental health resources and services.
• **Addressing the impact of gender-specific stressors**: Women may face unique stressors and challenges, such as gender-based violence, reproductive health issues, and workplace discrimination that can negatively impact their mental health.
• **Promoting gender equality**: Improving women's mental health is essential for promoting gender equality and addressing the systemic and cultural factors that contribute to poor mental health in women.

In summary, continued efforts to improve women's mental health through AI and other means are essential for promoting the wellbeing and equality of women and addressing the significant public health issue of mental illness.

12.13 Conclusion

The role of AI and ML in improving women's mental health is increasingly being recognized as a promising area of research and innovation. There are many potential applications of these technologies in this field, including the development of personalized treatment plans, early detection of mental health disorders, and the creation of virtual support systems. The capacity of AI and ML to sift massive volumes of data and recognize patterns that would be challenging or impossible for humans to find is one of the primary advantages of these technologies for mental health. This can lead to more accurate diagnoses and treatment recommendations, as well as earlier detection of mental health issues that may not be apparent through traditional screening methods. Another potential application of AI and ML in women's mental health is the development of virtual support systems, which can provide 24/7 access to resources and guidance. These systems can be particularly helpful for women who live in remote or underserved areas, as well as those who may be hesitant to seek help in person due to stigma or other barriers. Overall, while there is still much research to be done in this area, the potential benefits of AI and ML in improving women's mental health are significant. By leveraging the power of these technologies, we may be able to develop more effective, personalized treatments and support systems that can help women overcome mental health challenges and live happier, healthier lives.

References

[1] M. Iglesias, et al., Evaluating a digital mental health intervention (Wysa) for workers' compensation claimants: pilot feasibility study, J. Occup. Environ. Med. (2023), https://doi.org/10.1097/JOM.0000000000002762.

[2] C. Aldrich, H. Hanif, K. Parks, Normal Life Stress, they Say, Psychiatr, Times, 2022.

[3] F. Abdi, F.A. Rahnemaei, P. Shojaei, F. Afsahi, Z. Mahmoodi, Social determinants of mental health of women living in slum: a systematic review, Obstet. Gynecol. Sci. (2021), https://doi.org/10.5468/ogs.20264.

[4] A. Matsumoto, C. Santelices, A.K. Lincoln, Perceived stigma, discrimination and mental health among women in publicly funded substance abuse treatment, Stigma Heal. (2021), https://doi.org/10.1037/sah0000226.

[5] T.N. Salameh, L.A. Hall, T.N. Crawford, R.R. Staten, M.T. Hall, Likelihood of mental health and substance use treatment receipt among pregnant women in the USA, Int. J. Ment. Health Addict. (2021), https://doi.org/10.1007/s11469-020-00247-7.

[6] P. Grussu, I. Lega, R.M. Quatraro, S. Donati, Perinatal mental health around the world: priorities for research and service development in Italy, BJPsych Int. (2020), https://doi.org/10.1192/bji.2019.31.

[7] B.A. Brand, J.N. de Boer, P. Dazzan, I.E. Sommer, Towards better care for women with schizophrenia-spectrum disorders, Lancet Psychiatry (2022), https://doi.org/10.1016/S2215-0366(21)00383-7.

[8] R. Qasrawi, et al., Machine learning techniques for predicting depression and anxiety in pregnant and postpartum women during the COVID-19 pandemic: a cross-sectional regional study, F1000Research (2022), https://doi.org/10.12688/f1000research.110090.1.

[9] B.T. Sanford, et al., Tobacco treatment outcomes for hospital patients with and without mental health diagnoses, Front. Psychiatry (2022), https://doi.org/10.3389/fpsyt.2022.853001.

[10] C. Su, Z. Xu, J. Pathak, F. Wang, Deep learning in mental health outcome research: a scoping review, Transl. Psychiatry (2020), https://doi.org/10.1038/s41398-020-0780-3.

[11] O. Oyeniyi, S.S. Dhandhukia, A. Sen, K.K. Fletcher, A study of artificial intelligence frameworks and their capability to diagnose major depressive disorder, in: Lecture Notes in Computer Science (including subseries Lecture Notes in Artificial Intelligence and Lecture Notes in Bioinformatics), 2022, https://doi.org/10.1007/978-3-031-14135-5_1.

[12] S. Tutun, et al., An AI-based decision support system for predicting mental health disorders, Inf. Syst. Front. (2023), https://doi.org/10.1007/s10796-022-10282-5.

[13] E. Sajno, S. Bartolotta, C. Tuena, P. Cipresso, E. Pedroli, G. Riva, Machine learning in bio-signals processing for mental health: a narrative review, Front. Psychol. (2023), https://doi.org/10.3389/fpsyg.2022.1066317.

[14] S.E. Blackwell, T. Heidenreich, Cognitive behavior therapy at the crossroads, Int. J. Cogn. Ther. (2021), https://doi.org/10.1007/s41811-021-00104-y.

[15] S. Rachman, The evolution of behaviour therapy and cognitive behaviour therapy, Behav. Res. Ther. (2015), https://doi.org/10.1016/j.brat.2014.10.006.

[16] Y. Cheng, H. Jiang, AI-powered mental health chatbots: examining users' motivations, active communicative action and engagement after mass-shooting disasters, J. Contingencies Cris. Manag. (2020), https://doi.org/10.1111/1468-5973.12319.

[17] A. Bartlett, S. Hollins, Challenges and mental health needs of women in prison, Br. J. Psychiatry (2018), https://doi.org/10.1192/bjp.2017.42.

[18] V.S. Sakalidis, et al., Wellbeing of breastfeeding women in australia and new zealand during the covid-19 pandemic: A cross-sectional study, Nutrients (2021), https://doi.org/10.3390/nu13061831.

[19] A. Opanasenko, H. Lugova, A.A. Mon, O. Ivanko, Mental health impact of gender-based violence amid covid-19 pandemic: a review, Bangladesh J. Med. Sci. (2021), https://doi.org/10.3329/BJMS.V20I5.55396.

[20] S. Ali, D. Elsayed, S. Elahi, B. Zia, R. Awaad, Predicting rejection attitudes toward utilizing formal mental health services in Muslim women in the US: results from the Muslims' perceptions and attitudes to mental health study, Int. J. Soc. Psychiatry (2022), https://doi.org/10.1177/00207640211001084.

[21] J.A. Peterson, G. Chesbro, R. Larson, D. Larson, C.D. Black, Short-term analysis (8 weeks) of social distancing and isolation on mental health and physical activity behavior during COVID-19, Front. Psychol. (2021), https://doi.org/10.3389/fpsyg.2021.652086.

[22] C. Pieh, S. Budimir, T. Probst, The effect of age, gender, income, work, and physical activity on mental health during coronavirus disease (COVID-19) lockdown in Austria, J. Psychosom. Res. (2020), https://doi.org/10.1016/j.jpsychores.2020.110186.

Chapter 13

Artificial intelligence and machine learning for early-stage breast cancer diagnosis in women using vision transformers

S. Naveen Venkatesh[a], V. Sugumaran[a], and S. Divya[b]

[a]School of Mechanical Engineering (SMEC), Vellore Institute of Technology, Chennai, India,
[b]Centre for Advanced Data Science, Vellore Institute of Technology, Chennai, India

13.1 Introduction

Indian women have been increasingly suffering from specific health issues more than men in the recent years. Cervical cancer, breast cancer, stroke, polycystic ovarian disease (PCOD), polycystic ovarian syndrome (PCOS), and urinary tract infections (UTIs) are the key factors impacting quality of women's health in India. Breast cancer, which was the fourth most common cancer in Indian women in the previous decades, has now become the most prevalent form of cancer in the country. Breast cancer is a leading cause of death amongst women worldwide. Early detection and treatment greatly improve survival rates, and machine learning (ML) is proving to be a powerful tool in detecting breast cancer. Globally, the incidence of breast cancer has increased exponentially with an incidence rate of 39.1% amongst Indian women between 1990 and 2016 [1]. There are significant differences in terms of knowledge, education, treatment affordability, access to health care, and general perspective regarding breast cancer due to Indian society's diversified nature. As reported by current trends, Indian women show signs of developing breast cancer more commonly and at an earlier age than Western women [2]. This is alarming on various levels, as treatment plans and pain management become challenging as the disease progresses.

A disheartening fact is that more than 60% of women suffering from breast cancer discover their illness when they are in the third or fourth stage of disease

Artificial Intelligence and Machine Learning
for Women's Health Issues
https://doi.org/10.1016/B978-0-443-21889-7.00005-1

progression, when the symptoms can no longer be neglected [3]. Unfortunately, even healthcare workers were found to be largely unaware and lacking in knowledge about breast cancer symptoms and early detection signs [4]. A collective awareness, early symptomatic identification, and appropriate treatment plans would go a long way in arresting the increasing fatality rates and disease progression. A challenge evident in breast cancer patients is that visible symptoms can easily be overlooked as common everyday fatigue. In such situations, early detection would be effective in reducing mortality rates [5].

Two prominent methods of breast cancer detection are during a routine medical screening or when women experience pain or feel a lump [6]. With the rapid technological progress in ML algorithms, early diagnosis could be facilitated from medical imaging of previous instances of breast cancer and identifying leading patterns. For example, ML algorithms have been developed to diagnose and predict the prognosis of various types of cancers, including caner of the prostate and lungs and leukemia [7]. Traditionally, cancer in patients would be diagnosed using a combination of three tests: clinical examination, radiological imaging, and pathology tests, which form the foundation of conventional cancer diagnosis methods.

In instances of similar medical diagnosis, ML techniques have been deployed to assess images to help extract, classify, and predict illness [8]. As opposed to existing computer algorithms that run on preprogrammed algorithms and models, ML techniques enable computers to self-learn from existing information. They are built on recognizing patterns in the observed data and construct models anticipating the outcomes [9]. ML algorithms can identify patterns in large datasets that are beyond the capabilities of human analysis. With the increasing availability of medical imaging data, ML models are being developed to analyze mammography and other imaging modalities for the early detection of breast cancer. Recent advances in ML, such as vision transformers, have shown promising results in detecting breast cancer.

Medical images range from simple X-rays to CT scans, MRI scans, and in the present scenario of breast cancer, mammography images. As delay in diagnosis and treatment could be fatal for breast cancer patients, there is a critical need to create precise algorithms for classifying and recognizing tumors (cancerous and noncancerous) to advocate customized treatment without delay [10]. Competent risk prediction using modeling can assist radiologists in identifying high-risk female patients as well as setting up private screenings for women and empowering them to join in programs for early detection. Popular ML algorithms used for breast cancer prediction include artificial neural network (ANN), logistics regression (LR), k-nearest neighbor (KNN), decision tree, Naïve bayes algorithm, support vector machine (SVM), random forest, and so on. [11–14].

In the field of medical imaging diagnostics, especially for imaging breast cancer via mammography, InceptionV3, DenseNet121, ResNet50, VGG16, and MobileNetV2 models have been implemented to study deep convolutional neural networks (CNNs) for classifying mammogram images [15]. These computer-aided diagnostic (CAD) techniques have been improved in recent years to improve detection and classification of breast cancer amongst Indian women. These techniques have proven to have an efficacy almost on par with radiology experts in detecting breast cancer from mammographs [16,17].

Vision transformers are a type of deep learning (DL) model that can learn to recognize patterns in images without the need for handcrafted features. These models have shown superior performance compared to traditional CNNs in several computer vision tasks, including medical image analysis. ML algorithms can also be used to predict the risk of breast cancer in patients. By analyzing patient data, including age, family history, and other risk factors, ML models can predict the likelihood of developing breast cancer. This can help healthcare professionals identify patients who may benefit from increased screening or preventive measures [18]. In addition to improving detection and risk prediction, ML can also aid in the development of personalized treatment plans for breast cancer patients.

The development of unique artificial intelligence (AI) techniques and new image collection methods has led to significant breakthroughs in cancer screening in recent years. Although AI has been improving algorithms to perform more accurate and faster diagnostics related to breast cancer screening, its ability to counter the subjective character and increase the effectiveness of human picture interpretation is in its nascent stages. DL techniques have compounded the field of medical AI research, causing medical practitioners, AI developers, and medical personnel to be hopeful about the future of cancer care. Most of the DL or human-engineered AI systems act as supplementary medical assistants to radiologists for explicit tasks [19].

In the present study, vision transformer (VT), a type of CNN technique, is built and tested for breast cancer diagnosis in Indian women. Being a recently created and prominent CNN model, the VT has drawn considerable interest in a variety of medical diagnostic applications [20]. Validating the VT's impact on imaging-based breast cancer detection is the need of the hour to improve health care. In several classification benchmarks, the VT model demonstrated greater efficacy than other CNN models, although its use in breast cancer detection has been sparse. This gap in technical methodology has been addressed in the present study by constructing a VT model to detect and classify breast cancer from mammographs of Indian women.

13.2 Vision transformer

Researchers looked at a transformer's potential in computer vision after observing its high performance in natural language processing (NLP)-related tasks. One of the early initiatives of this observation was the VT. Recent image classification challenges have shown amazing results for VTs, which have several benefits over traditional CNNs. These benefits include long-range connections, flexible modeling, and attention maps that provide insight into what the model prioritizes inside an image [21].

The use of VTs in common image recognition tasks including object identification, image segmentation, picture classification, and action recognition. Additionally, VTs are used in multimodel tasks such as visual reasoning, visual question answering, and generative modeling. VTs allow models to learn image structure independently since pictures are represented as sequences and class labels for the image are anticipated. Input pictures are handled as a series of patches, with each patch being "flattened into a single vector by concatenating the channels of all of its pixels" before being linearly projected to the specified input dimension.

The effectiveness and scalability of VT is attributed to its three-tier architecture, which is made up of several flattened (two-dimensional) image patches, an encoder, and a classification head. The transformer's original structure is retained by the VT. However, to account for the transformer's input requirement, the medical image is broken down into a collection of image patches. This image-patch embedding process is basically a linear transformation, which in turn is a fully connected (FC) layer. In VT, the input image, which is the mammograph, is partitioned into 16×16 pixel squares. Spatial position of each patch is provided by a "positional encoding method" that varies from 1 to n, where in the position of each patch is encoded [9]. These individual "splint-patches" are then consecutively fed through the embedding layer and output vectors.

This encoder is made up of numerous blocks, with each block consisting of three crucial processing components: layer norm, multihead attention network (MHANet), and multilayer perceptron (MLP). When a picture is entered into VT, two processes emerge sequentially: embedding the individual split-image patches followed by positional encoding. The input picture, which consists of the dimensions height (H), width (W), and channel count (C), is broken into two-dimensional patches of reduced dimensions to arrange the input image data similarly the NLP domain's input structure. The number of patches is thus calculated from these dimensions.

The formula used to calculate the number of patches is:

$$\text{Number of patches} = (\text{image\_size}//\text{patch\_size}) ** 2 = \frac{HW}{P^2} \qquad (13.1)$$

Changing to two-dimensional structure, sequential imbedding, learnable embedding followed by patch embedding would be performed before the image patches pass through the transformer's encoder. The smaller the patch size, the greater the performance and the computational cost. Thus, a 16×16 patch size would be adopted, owing to its resilience against performance deviation and complex computational algorithms.

13.3 Methodology

The present study was conducted in four phases: (1) collecting data, (2) preprocessing data, (3) building model, and (4) model evaluation.

13.3.1 Dataset description

In this study, we propose a transfer-learning model built upon VT architecture for early detection of breast cancer. This model has been trained and tested using data collected from a few publicly available datasets, including INBreast data, Mammographic Image Analysis Society (MIAS) data, and Digital Database for Screening Mammography (DDSM) data. The data collected from these databases consisted of around 25,000 mammograph images. The DDSM, which is an open dataset collection, has around 2600 mammographic images containing normal, benign, and malignant cases, each of them tagged with validated pathological information [22]. The DMIAS is another database dedicated to mammograph images, with certain modifications making all the images the same pixel size (1024 × 1024). The INBreast database is a "mammograph-specific" database, consisting of around 115 patient mammographs, 90 of which belong to women with breast cancer and 25 of which belong to mastectomy patients. The dataset also includes mammographs with various types of lesions and distortions, such as masses, calcifications, and asymmetries [23]. The consolidated dataset includes 13,710 mammographs with malignant masses and 10,866 with benign masses. Postdata collection, the second step of data preprocessing, is essential. The image data set must be preprocessed before it can be used as a model input. For instance, the fully connected layers of CNNs require that all the pictures to be processed effectively for superior classification. Fig. 13.1, presents the sample mammograph images of malignant and benign breast cancer that need to be preprocessed. These are sample images representing the collated dataset Fig. 13.2 presents the architecture depicting the overall workflow of VTs.

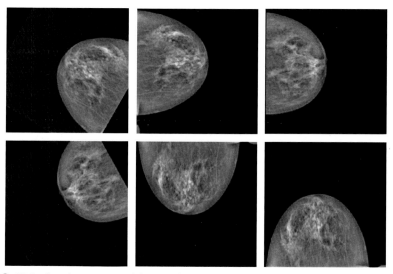

FIG. 13.1   Sample mammograph images of malignant and benign breast cancer.

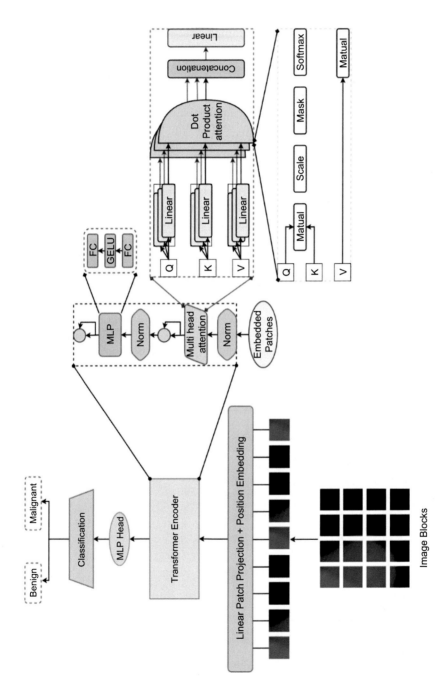

FIG. 13.2 VT workflow process.

FIG. 13.3   Image input to VT.

Fig. 13.3 shows the processed image that would be input to the VT. Apart from transformer layers, self-attention is also used in VTs to analyze data. A crucial part of the VT architecture is the self-attention mechanism, used for interpixel interaction (attention) and to extract contextual information (attention scores) from the input pictorial data. Due to their relevance to the job at hand, the VT model focuses on various areas of the input data. The input data are calculated as a weighted sum, with the weights determined by how similar the input characteristics are. This enables the model to give the pertinent input characteristics greater weight, which aids in capturing more accurate representations of the input data. A network may learn the hierarchies and alignments contained in incoming data by using this to measure paired entity interactions.

VT does not employ convolutional layers, in contrast to CNNs. As an alternative, it separates the input picture into a series of patches. These patches run through a succession of transformer layers. Each transformer layer focuses on understanding various aspects of the picture, such as color, shape, and texture. Eventually, a classification layer is designed to receive the output from the transformer layers and, later, a label is applied to the image. VT may be used for many different things, such as segmentation, object identification, and picture categorization. It could also serve as the foundational infrastructure for later activities like image synthesis, picture-to-image translation, and super resolution.

The hyper parameters used to train VT model include learning rate, weight decay, batch size, number of epochs, image size (where the images are resized), patch size (size of each patch that need to be extracted from the input images), number of patches, projection dimensions, number of heads, transformer layers, and MLP head units (size of the dense layers present in the final classifier). There is an 80–20 split in the training and testing datasets. VT has an innovative architecture for computer vision tasks. It is built using a DL approach for NLP called the transformer architecture. It has been modified for tasks like object identification and picture categorization. The transformer blocks used by VT are made up of a point-wise feed-forward layer and a multihead self-attention layer. The model can learn the long-range relationships between the components of the input picture thanks to the multihead self-attention layer. The output from the self-attention layer is processed by the point-wise feed-forward layer. VT excels in picture classification and object recognition tasks thanks to its ability to learn from huge datasets.

13.4   CNN-based breast cancer detection—Vision transformer

The overall dataset was split into two subsets comprising training data (19,649 images) and validation or testing data (4947 images). To automate the separation of training and test data, a randomized dataset distribution was adopted. To ensure that the findings of the categorization are objective, a homogenous dataset is preferable. Uneven datasets, lacking in uniformity, might lead to bias during categorization in favor of a certain condition or class. More datasets and increased training time must be facilitated when using DL-based approaches so that the learning rate is increased. Training and validation steps were performed using Google's Colaboratory platform.

13.4.1   Training process and model evaluation

The collected data must be preprocessed so that an effective training and testing dataset can be created. A dataset's overall dimensionality is decreased, and extra training examples (pictures) are removed as part of the instance selection step, an effective way of preprocessing the collected data. To train VT, some hyperparameters were used as settings for the model building. Table 13.1 lists the training setting for each of the hyperparameters.

Fig. 13.4 presents the performance of the proposed VT model, specifically the accuracy and number of iterations. Plots demonstrate how the proposed CNN-based VT may train to a point of rapid convergence for the testing dataset. The model also shows little inaccuracy, demonstrating the network's ability to generalize. The validation dataset was chosen at random, therefore it is clear that the model tries to match the validation dataset more closely when compared to the training dataset. This predicament arises because of the built-model's dropouts. It is also apparent that as the number of iterations grows, the model's

TABLE 13.1 Hyperparameters for training VT.

Hyperparameter	Value
Learning rate	0.0001
Weight decay	0.0001
Batch size	8
Epochs	100
Input image size	72
Patch Size	16
Projection dimension	64
MLP head units	[2048, 1024]
Transformer layers	8
Number of heads	4

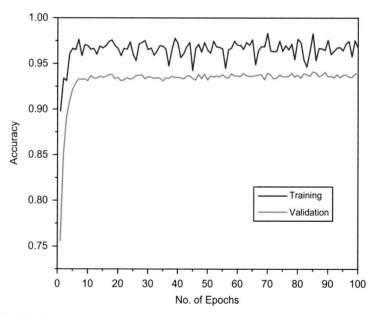

FIG. 13.4 Overall accuracy of the model for the dataset.

classification accuracy improves. Model efficacy is claimed to be better when underfitting is avoided. This is a key metric to monitor during model training. Underfitting and extensive errors would be incurred if the constructed model fails to effectively represent the training set of data. The high accuracy of the developed model will be a product of least validation loss. If this is not addressed, it might lead to the generation of errors.

The plot shown in Fig. 13.4 indicates the smooth training and validation processes in the dataset. Around 100 iterations were performed throughout the training and validation processes. One can observe from Fig. 13.4 that the training progress curve reached above 95% and maintained a pattern that oscillates between 93% and 98%. This shows that the training is uniform, and that the VT has learned the image features accurately. The overall accuracy for the dataset was found to be 93.79%. After the model accuracy is deduced, a confusion matrix is constructed. This table is used to assess the efficiency of the model with respect to its class-categorization. A summary of the classification values is tabulated in this matrix with the real values and the predicted values. The accuracy of the CNN model presented in this study is evaluated using the confusion matrix.

In a typical CNN-based confusion matrix, the rows represent the actual classes and the columns represent the predicted classes. The confusion matrix displays the total number of "true positives, true negatives, false positives, and false negatives" for each class. The number of samples that were properly identified for each class is represented by the confusion matrix's diagonal elements. The total number of incorrectly identified samples may be calculated by adding together all the off-diagonal components ("false positives and false negatives"). The number of samples that were properly identified is divided by the sum total of samples to determine the model's accuracy. By modifying the model's hyperparameters (Table 13.1), the accuracy of the model could be further improved. Model accuracy is optimized to further decrease the instances of "false positives" and/or "false negatives."

Fig. 13.5 presents the confusion matrix. The accuracy of classification is depicted in comparison to the predicted values by displaying in a matrix form. The row represents the actual benign and malignant cases present in the mammograph dataset. The vertical axis presents the predicted cases.

Based on the confusion matrix shown in Fig. 13.5, out of a total of 203 malignant cases, around 104 were correctly diagnosed. Out of the 2644 benign cases, 1966 were correctly diagnosed. The confusion matrix shows the two classes: benign and malignant images. The confusion matrix in Fig. 13.5 is used to calculate each condition's specific classification accuracy. The classification accuracy is determined by the ratio of accurately predicted classes from the mammographs to the sum total images taken into account for each class classification. We obtained overall test accuracy of 93.79% for the proposed model in this study.

13.4.2 Evaluation of CNN models in diagnosing breast cancer from mammographs

Several research studies have been conducted to diagnose breast cancer from mammographs using various CNN models. Table 13.2 provides a consolidated

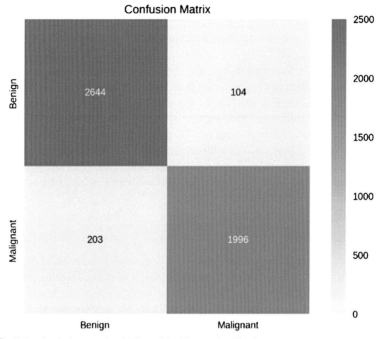

FIG. 13.5   Confusion matrix of VT model with two classifications.

TABLE 13.2 Accuracy comparison of CNN models in detecting breast cancer from mammographs.

S·No	CNN model	Dataset and sample size	Accuracy	Author
1	BCD CNN	332 mammographs	82.71%	[27]
2	AlexNet	332 mammographs	91.4%	[24]
3	MRFE-CNN	DDSM mammographs	87.1%	[26]
4	CNN	322 mammographs	92.1%	[25]
5	VT	25,000 mammographs	93.79%	Present study

analysis, comparing the accuracy of CNN models built to detect breast cancer. CNN techniques like Block Coordinate Descent (BCD), AlexNet (an eight-layer deep image classification model), Multi-Route Feature Extraction (MRFE) were compared with the VT used in this study. [24–27]. These models have been built and tested by researchers to produce competent results in segregating images as per their defined classes. From the numerical results presented in Table 13.2, it can be deduced that the proposed solution provides greater accuracy in terms of classification (93.79%) compared to the other models.

Table 13.2 lists the comparative data from similar studies on the accuracy levels obtained when CNN models are built to detect breast cancer from mammograms. It could be observed that the present study, which has implemented a VT-based CNN technique, has greater accuracy when compared to other models. The results obtained using the VT technique prove that this technique can improve model performance with accuracy of 93.79%. The performance of modeling can be improved by adopting different ML techniques and being able to access larger datasets from several institutions while considering key attributes from various pertinent data sources.

13.5 Discussion

The results show that the mammography diagnostic and detection system proposed in this study based on the VT model garnered an accuracy rate of 93.79% on the validation data set. This is an improved accuracy when compared to the other models discussed in the study [5,20,21,23,28,29]. The improved accuracy is owed to the usage of the VT model to classify the dataset. In this chapter, we established a VT-based classification system for detecting breast cancer from mammography images. The usage of the collated dataset also contributed to improving the model training and hence the model accuracy. Future studies could improve on a few issues. For example, the model could only detect benign and malignant images from the mammographs. Using uniform datasets would improve model training.

13.6 Conclusion

Incidence and mortality rates of breast cancer in Indian women are increasing with each passing year. Creating public awareness of breast cancer starts with acknowledging that it is now one of the most prevalent neoplasms amongst Indian women. It is rapidly evolving into a public health catastrophe, which could be tackled by advocating and adopting early detection techniques. Capitalizing on medical imaging techniques using DL methodologies, this chapter presented a robust DL technique. For women exhibiting breast cancer signs, the applications of the ML algorithm presented in this study would help with quicker prediction, precise diagnosis, and early intervention. The study's test findings show that VTs are more accurate when analyzing mammography pictures in clinical situations. With the purpose of enhancing diagnostic breast cancer research, this study may be expanded in the future to include a variety of datasets and classification criteria. There is scope for improving algorithmic accuracy that could be addressed while developing DL algorithms in the future.

Researchers could use several data augmentation strategies to address the problem of small datasets. The prospect of any disparity between positive and negative data, which could result in bias towards either a positive or negative prediction, should be considered in future studies. For accurate diagnosis and breast cancer prognosis, the disparity in the amount of breast cancer pictures versus afflicted patches must be addressed. The VT algorithm presented in this chapter was able to accurately classify breast cancer incidence. This ML technique proves to be a viable option that could improve early detection of breast cancer in Indian women. Adopting these technologies would improve the present screening programs and treatment options available, leading to a favorable clinical outlook in the country.

References

[1] R. Mehrotra, K. Yadav, Breast cancer in India: present scenario and the challenges ahead, World J. Clin. Oncol. 13 (3) (2022) 209.

[2] A.P. Maurya, S. Brahmachari, Current status of breast cancer management in India, Indian J. Surg. 83 (Suppl 2) (2021) 316–321.

[3] A. Gogia, S.V.S. Deo, D. Sharma, S. Mathur, Breast cancer: The Indian Scenario, 2020.

[4] V. Fotedar, R.K. Seam, M.K. Gupta, M. Gupta, S. Vats, S. Verma, Knowledge of risk factors & early detection methods and practices towards breast cancer among nurses in Indira Gandhi medical college, Shimla, Himachal Pradesh, India, Asian Pac. J. Cancer Prev. 14 (1) (2013) 117–120.

[5] E.H. Houssein, M.M. Emam, A.A. Ali, P.N. Suganthan, Deep and machine learning techniques for medical imaging-based breast cancer: a comprehensive review, Expert Syst. Appl. 167 (2021) 114161.

[6] J. Arevalo, F.A. González, R. Ramos-Pollán, J.L. Oliveira, M.A.G. Lopez, Representation learning for mammography mass lesion classification with convolutional neural networks, Comput. Methods Prog. Biomed. 127 (2016) 248–257.

[7] M.K. Gupta, P. Chandra, A comprehensive survey of data mining, Int. J. Inf. Technol. (2020) 1–15. Feb.

[8] M.M. Islam, M.R. Haque, H. Iqbal, M.M. Hasan, M. Hasan, M.N. Kabir, Breast cancer prediction: a comparative study using machine learning techniques, SN Comput. Sci. 1 (2020) 1–14.

[9] M. Cantone, C. Marrocco, F. Tortorella, A. Bria, Convolutional networks and transformers for mammography classification: an experimental study, Sensors 23 (3) (2023) 1229.

[10] R. Rabiei, S.M. Ayyoubzadeh, S. Sohrabei, M. Esmaeili, A. Atashi, Prediction of breast Cancer using machine learning approaches, J. Biomed. Phys. Eng. 12 (3) (2022) 297–308. 10.31661/jbpe.v0i0.2109-1403. PMID: 35698545; PMCID: PMC9175124.

[11] M. Abdar, M. Zomorodi-Moghadam, X. Zhou, R. Gururajan, X. Tao, P.D. Barua, R. Gururajan, A new nested ensemble technique for automated diagnosis of breast cancer, Pattern Recogn. Lett. 132 (2020) 123–131.

[12] K. Adem, Diagnosis of breast cancer with stacked autoencoder and subspace kNN, Physica A Stat. Mech. Appl. 551 (2020) 124591.

[13] Z. Khandezamin, M. Naderan, M.J. Rashti, Detection and classification of breast cancer using logistic regression feature selection and GMDH classifier, J. Biomed. Inform. 111 (2020) 103591.

[14] S. Punitha, F. Al-Turjman, T. Stephan, An automated breast cancer diagnosis using feature selection and parameter optimization in ANN, Comput. Electr. Eng. 90 (2021) 106958.

[15] H. Tabrizchi, M. Tabrizchi, H. Tabrizchi, Breast cancer diagnosis using a multi-verse optimizer-based gradient boosting decision tree, SN Appl. Sci. 2 (2020) 1-19.

[16] W.M. Salama, M.H. Aly, Deep learning in mammography images segmentation and classification: automated CNN approach, Alex. Eng. J. 60 (5) (2021) 4701–4709.

[17] J.G. Melekoodappattu, A.S. Dhas, B.K. Kandathil, K.S. Adarsh, Breast cancer detection in mammogram: combining modified CNN and texture feature based approach, J. Ambient. Intell. Humaniz. Comput. (2022) 1–10.

[18] L. Wang, Y. Zhu, L. Wu, Y. Zhuang, J. Zeng, F. Zhou, Classification of chemotherapy-related subjective cognitive complaints in breast Cancer using brain functional connectivity and activity: a machine learning analysis, J. Clin. Med. 11 (8) (2022) 2267.

[19] J. Xiao, M. Mo, Z. Wang, C. Zhou, J. Shen, J. Yuan, Y. Zheng, The application and comparison of machine learning models for the prediction of breast cancer prognosis: retrospective cohort study, JMIR Med. Inform. 10 (2) (2022) e33440.

[20] A. Dosovitskiy, L. Beyer, A. Kolesnikov, D. Weissenborn, X. Zhai, T. Unterthiner, N. Houlsby, An Image is Worth 16x16 Words: Transformers for Image Recognition at Scale, 2020. arXiv preprint arXiv:2010.11929.

[21] S. Tuli, I. Dasgupta, E. Grant, T.L. Griffiths, Are Convolutional Neural Networks or Transformers More Like Human Vision?, 2021. arXiv preprint arXiv:2105.07197.

[22] R.S. Lee, F. Gimenez, A. Hoogi, K.K. Miyake, M. Gorovoy, D.L. Rubin, A curated mammography data set for use in computer-aided detection and diagnosis research, Sci. Data 4 (2017) 170177, https://doi.org/10.1038/sdata.2017.177.

[23] I.C. Moreira, I. Amaral, I. Domingues, A. Cardoso, M.J. Cardoso, J.S. Cardoso, INbreast: toward a full-field digital mammographic database, Acad. Radiol. 19 (2) (2012) 236–248.

[24] L. Dong, Diagnosis of breast cancer from mammogram images based on cnn, Journal of the Institute of Industrial Applications Engineers 8 (4) (2020) 117–121.

[25] S.Z. Ramadan, Using convolutional neural network with cheat sheet and data augmentation to detect breast cancer in mammograms, Comput. Math. Methods Med. 2020 (2020).

[26] R. Ranjbarzadeh, N. Tataei Sarshar, S. Jafarzadeh Ghoushchi, M. Saleh Esfahani, M. Parhizkar, Y. Pourasad, M. Bendechache, MRFE-CNN: multi-route feature extraction model for breast tumor segmentation in mammograms using a convolutional neural network, Ann. Oper. Res. (2022) 1–22.

[27] Y.J. Tan, K.S. Sim, F.F. Ting, Breast cancer detection using convolutional neural networks for mammogram imaging system, in: 2017 International Conference on Robotics, Automation and Sciences (ICORAS), IEEE, 2017, pp. 1–5.

[28] M. Gupta, H. Wu, S. Arora, A. Gupta, G. Chaudhary, Q. Hua, Gene mutation classification through text evidence facilitating cancer tumour detection, J. Healthc. Eng. 2021 (2021) 1–16.

[29] S. Kour, R. Kumar, M. Gupta, Study on detection of breast cancer using machine learning, in: 2021 International Conference in Advances in Power, Signal, and Information Technology (APSIT), IEEE, 2021, October, pp. 1–9.

Chapter 14

Recent and future applications of artificial intelligence in obstetric ultrasound examination

Shalu Verma[a], Alka Singh[b], Kiran Dobhal[c], Nidhi Gairola[a], and Vikash Jakhmola[a]

[a]*Uttaranchal Institute of Pharmaceutical Sciences, Uttaranchal University, Dehradun, Uttarakhand, India,* [b]*School of Pharmaceutical Sciences, Sardar Bhagwan Singh University, Dehradun, Uttarakhand,* [c]*College of Pharmacy, Shivalik Campus, Dehradun, Uttarakhand*

14.1 Introduction

Ultrasound is used throughout the entirety of pregnancy. It is necessary for both the identification and treatment of illnesses as well as the observation of fetal growth and development. It can deliver high-quality photos and more accurate diagnostic information together with extensive information on fetal anatomy. Accurate diagnosis and treatment depend on having access to high-quality obstetric ultrasound imaging [1]. However, ultrasound during the first trimester of pregnancy is easily affected by involuntary fetal movements. In the second and third trimesters, the structures of interest are virtually always obscured, which can make inspection challenging and lead to incorrect diagnoses. The acquisition of high-quality ultrasound pictures, standardized measures, and accurate diagnosis are all impacted by measurement and observation subjectivity [2].

Diagnostic radiography is one area of medicine where artificial intelligence (AI) is becoming more and more popular, and its application has been expanding to encompass a wide range of imaging-based diagnoses. As humans have been diagnosing certain diseases for a very long time, many AI techniques have been utilized to enhance the diagnosis process. In addition to helping doctors diagnose patients, these techniques both directly and indirectly lower

Artificial Intelligence *and* Machine Learning
for Women's Health Issues
https://doi.org/10.1016/B978-0-443-21889-7.00018-X

medico-legal difficulties. AI-based automatic measures and evaluations have been implemented in the past 10 years to decrease interobserver and intrameasurement variability to increase diagnosis precision. Structure recognition, standard and automatic measurements, and classification diagnosis are the three components of AI in improving obstetric ultrasonography accuracy, as shown in Fig. 14.1 [4]. Because obstetric ultrasonography takes a lot of time, using AI could speed up workflow and cut down on examination times. Most of the relevant studies remain in development, even though much obstetric ultrasonography with high-resolution images and precise measurement, commercial software, and AI-assisted approaches have been established [5]. One of the most important elements to proving the reliability and clinical applicability of AI algorithms is the promotion of interdisciplinary collaboration. Here, we thoroughly examine and assess the use of AI in obstetric ultrasound, as well as its benefits and drawbacks [6]. AI-assisted obstetric ultrasound can be categorized according to pregnancy trimester [7]. Virtual reality (VR) learning and telemedicine or telediagnosis services are some prospects for an obstetric ultrasound that warrant attention. We think that interdisciplinary research teams working together will make it easier to translate algorithms into real-world clinical applications for AI-assisted obstetric ultrasound [3].

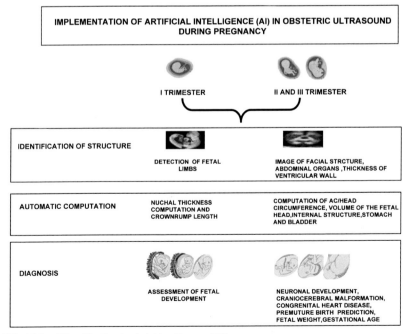

FIG. 14.1 AI's utilization in obstetric ultrasound [3].

14.1.1 Prepregnancy scan

Transvaginal ultrasonography is crucial in determining early pregnancy, whether a pregnancy is ectopic, or if there has been a miscarriage. Using transvaginal sonography, measuring the crown-rump length and average sac diameter carefully, and adopting acceptable value cutoffs for these parameters are necessary to prevent making a false-positive miscarriage diagnosis. A previous study found that the lack of an intact embryo in the amniotic sac is a good indicator of miscarriage. In contrast to the appearance of an irregular adnexal mass, a gestational sac outside the womb that contains an embryo or a yolk sac (also known as the "blob" sign) and heart activity is a symptom of ectopic pregnancy, while a structure resembling an extrauterine sac (also known as the "bagel" sign) strongly suggests a tubal ectopic pregnancy. Inadequate insertion of the uterus inside of a Caesarean scar or nearby, as well as the typical symptoms of disorders in the placenta accreta range, is ultrasonography findings of Caesarean scar pregnancy that should be watched for in women who have previously undergone this procedure [8].

14.2 Utilization of AI in the first trimester

14.2.1 Automated evaluation of fetal development and growth

It is essential to assess fetal growth and development during the first trimester to diagnose and treat pregnancy issues like preterm labor and low birth weight. The standard procedure for assessing fetal growth is measuring the length of the crown-rump by using two-dimensional ultrasonography (2D-US). The subjective reliance on single-section measurements by 2D-US, however, may not reveal any appreciable distinction in the first trimester between normal and abnormal fetuses in terms of the length of the crown to the rump. In comparison to 2D image measures, images in three dimensions volume measures may provide more details about fetal development, but because 3D imaging takes time, the volume measurements may not be accurate [9]. As a result, a method for calculating the volume of 3D images that is semiautomatic and based on point-of-interest identification and pixel extraction has been developed. This technique was used to identify fetal contours and compute fetal volume using a 12-week-old fetus, demonstrating the efficacy of imaging systems for precise identification. Even so, certain segmentation errors needed to be corrected manually because the semiautomatic system failed to detect irregular elements [10]. A new study employed a 3D convolutional neural network (CNN) method to gauge the volume of 104 fetuses and their extremities during the first trimester (10–14 weeks) [11]. This made it possible to segment the placenta, some anatomical components, and the gestational sac, which contained the fetus [12]. Another study put forth a 3D ultrasound-based image processing technique that included automated biometric measurements, fetal segmentation, identification of fetal limbs, and estimation of standard biometry views. Given the close relationship between the fetus and fetal appurtenances, these algorithms can result in a more methodical and thorough evaluation [13].

14.2.2 Nuchal translucency thickness assessment

For the early diagnosis of chromosomal abnormalities, 2D-US is essential for measuring nuchal translucency (NT) thickness. A typical median sagittal picture should be used to measure NT, which is the greatest thickness between the subcutaneous soft tissue at the cervical spine level and the skin of the fetus [14]. Doing so requires a high level of accuracy and knowledge. The assessment of NT thickness frequently necessitates repeated attempts because of frequent fetal movements, small fetal structures, and poor imaging quality [15]. In one study, the fetal NT thickness was assessed using semiautomated methods to minimize measurement errors and challenges, and the findings comparisons were made with mechanical measurements. Another study used deep belief networks to determine the fetal NT structure while measuring the fetal NT thickness using a standard median sagittal picture. Therefore, sagittal plane information and 3D picture data were combined to create an automatic recognition model, which had a detection accuracy of 88.6% [16].

14.3 Utilization of AI-assisted ultrasonography in the second and last trimesters

14.3.1 Fetal head

Prenatal ultrasonography is essential to check fetal growth and development, find fetal anomalies, and treat prenatal diseases to lower mortality rates [17]. One of the most difficult organs to view using prenatal ultrasound is the fetal brain. Fetal cranial capacity, head biometry, assessment, totally automatic segmentation of the fetus' internal anatomy, and classification of normal and abnormal ultrasound results are all done in the fetal brain using AI-aided ultrasonography [18].

14.3.2 Automated fetal head segmentation and inner composition

Head circumference (HC) is a key marker of fetal growth, and aberrant readings may point to a restriction in growth. This measurement, along with others like the biparietal diameter, can be used to calculate the week of pregnancy [19]. Moreover, prenatal ultrasound uses a variety of measurements and sections to effectively monitor craniocerebral development and identify intracranial abnormalities. Nevertheless, most underdeveloped nations do not have enough qualified sonographers, and novice sonographers may have trouble obtaining high-quality standard plane images [20]. This might have an impact on calculating the pregnancy week and assessing the growth and development of the fetus. The semiautomatic measurement software used by ultrasonic machines to measure HC requires the localization of two sites, usually for the short diameter locating point on the parietal diameter section, and may lead to measurement inaccuracy [21]. HC has previously been measured using a variety of conventional techniques, including active contouring and randomized Hough transform. Previous years have seen the investigation of novel techniques based on deep learning (DL) techniques [22]. Without the acquisition of a standard plane, one investigation employed the obstetric sweep routine to perform an entirely autonomous analysis. The approach was built on two complete CNNs, where HC could be predicted from earlier data and gestational age (GA) could be calculated using the Hadlock curve. The ability to partially avoid the inaccuracies caused by traditional plane scanning was a significant achievement. A CAD system that uses the random forest technique to extract HC automatically was reported in another study. The research, which was economical and appropriate for medical facilities without skilled sonographers, contained data from all trimesters, established the development curve, and confirmed that it was required to evaluate results for each trimester independently [23]. Yang et al. developed a fully automated technique to segment the entire fetal head based on 3D ultrasonography, which achieved a dice similarity coefficient of 96% to offer a basis for the extraction of representative biometrics in the fetal head. The research team also looked at a broad framework for automatically identifying fetal anatomical components in 2D-US pictures, which enabled the availability of objective biometric measurements [24]. It demonstrated exceptional skill in segmenting fetal HC and abdominal circumference (AC). In addition to HC, studies have focused on precisely segmenting the fetal head's interior structures. Fetal lateral ventricles, the cisterna magna, trans cerebellar structures, the posterior horn of the lateral ventricle, and other structures were frequently measured [25]. Except for the picture quality (the fetal head was farther from the ultrasound probe and protected by nearby tissue or anatomical structures) and the challenges of localization brought on by fetal activity, the approaches were not significantly different from those utilized for newborns. Most of these studies used DL algorithms or other types of commercial software as their foundation [26].

14.3.3   Cervical function assessment and prediction

The primary cause of spontaneous preterm births and repeated abortions in the second trimester is cervical insufficiency, which also contributes to newborn mortality. To assess cervical function during pregnancy, transvaginal ultrasonography is frequently performed [27]. There are usually too many tests performed and too many treatments administered because there aren't any useful methods for assessing cervical function.

The main use of AI in the assessment of cervical insufficiency is the assessment of tissue texture in images [28]. Spanish researchers used a quantitative examination of cervical roughness to gauge changes in cervical tissue during pregnancy. By feature transformation, data segmentation, and model computation, a prediction model of pregnancy week age based on features from the cervical picture was created, which demonstrated that there was a substantial association between cervical ultrasound images and pregnancy week age. For each ultrasound picture, 18 features in the area of interest were extracted [29]. The labium entries' center was chosen as the area of interest, the feature combination learning algorithm was put into practice, the CTx score was established, and it was determined that pregnant women with short cervixes who delivered their babies at full term had higher CTx scores than those who delivered them prematurely [30]. This was done because cervical length (CL) has a low specificity for assessing cervical function. With the use of this method, it was possible to forecast the likelihood of an early birth in expectant women with short cervixes [31].

The use of omics to anticipate preterm birth (PTB) has been made in addition to texture analysis. In a prior study, DL was utilized along with several machine learning (ML) techniques to predict preterm latency, PTB, and the length of time needed for neonates to receive treatment during the second trimester in the NICU. Despite the small sample size, the study discovered that when it came to handling complex data, DL outperformed other ML algorithms in the multifactorial prediction of cervical insufficiency [32].

14.3.4 Determination of fetal weight and gestational age

For extreme fetal weights, AI technology has revealed its potential to generate fetal weight accurately, which is primarily dependent on week of pregnancy and femur length, biparietal diameter, and the AC [33,34].

Given that preterm delivery is the primary cause of death for children younger than 5 years, it represents a significant global health concern. Existing methods for determining fetal GA are often imprecise. For instance, during 20–30 weeks of gestation, the 95% prediction interval is roughly estimated to 18–36 days [35]. To address this issue, Russell et al. created an ML method to precisely estimate GA utilizing fetal biometric and ultrasound-derived data. Instead of the mother's last menstrual cycle date, the accuracy of the approach was assessed to precisely gather data about each fetus, in particular, the intervals between ultrasound appointments. The data from a different and more varied population was used to demonstrate how the technique was generalized. Using measures taken over 10 weeks covering the second and third trimesters, the researchers calculated GA within 3 days between 20 and 30 weeks of gestation with a 95% CI. As a result, the prediction interval for estimating fetal GA in the 20–30-week GA window is three to five times better than it was for any previous methods [36]. This would make it feasible to better manage individual pregnancies and more precisely identify fetuses who are in danger than is currently possible. The increased precision was anticipated to enhance population-level fetal growth charts and population health evaluations [37,38]. Long-standing restrictions in predicting fetal GA and future growth trajectories could be overcome by ML without the use of frequently erroneous information that is already public, like the mother's latest menstrual cycle date [39]. The application of this approach in clinical settings may facilitate the management of pregnancies and improve population health [40].

14.4 Ultrasound by utilizing Doppler

In obstetrics, Doppler ultrasound is frequently utilized. The International Society of Ultrasound in Obstetrics and Gynecology (ISUOG) has provided advice on how to do fetoplacental circulation Doppler ultrasonography. Finding late-onset fetal growth restriction (FGR) is difficult. For fetuses that are appropriate for their gestational age (AGA) or undersized for their gestational age (SGA), the cerebroplacental ratio (CPR) is an essential measure for forecasting unfavorable pregnancy outcomes. The middle cerebral artery (MCA) pulsatility index (PI) is divided by the umbilical artery (UA) pulsatility index (PI) to determine CPR. A low prediction rate for a poor perinatal outcome was seen with CPR, whether it was adjusted for the centile of projected fetal weight or not. An aberrant uterine artery (UtA) Doppler in the third trimester may help predict neonatal death in fetuses that are thought to be undersized for GA. According to a recent prospective study, CPR and UtA Doppler alone were not as effective at detecting FGR as the cerebral–placental–uterine ratio (CPUR; CPR divided by mean UtA pulsatile index) [41].

14.5 Preterm birth

Early detection of PTB in the first and second trimesters would help to improve pregnancy outcomes. Research has focused on developing an effective PTB prediction model that relies on artificial neural networks (ANNs). Many studies suggested that PTB in the first trimester of pregnancy might be predicted using CL sonographic measures. Nonetheless, additional research is required to outline the CL's capacity to screen PTB. The advantages of DL and ML over the logistic regression approach are their capacity to handle highly dimensional patient data and their capacity for self-learning [42].

According to this application of AI, the key difficulty in other medical IVF streams is still choosing a viable embryo. This appears to be crucial in estimating the outcomes that would result in a shorter pregnancy period and live birth of a healthy child. IVF, which has sparked increasing interest and gained traction in the commercial sectors, would be automated, precise, and standardized, according to Zaninovic et al. Also, the role of AI in embryology has drawn considerable attention and exposed the validity of different fields of reproductive research [43]. Furthermore, AI-powered tactics have the potential to be quick, accurate, and important. The effectiveness of therapy and diagnosis of reproductive system illnesses will increase if AI is extensively used to evaluate patient features including endocrine status, ovarian reserve, age, and endocrine status. These parameters support the likely outcomes of successful IVF through AI support and tools. The challenges to achieving high sensitivity during pregnancy include various unidentified parameters that lead to positive IVF results, which are required to train AI. Larger datasets, including computer vision, were used for this statement to design the ANN model effectively and maximize its predictive power. Only a few other prior attempts using AI techniques to evaluate human oocytes, forecast typical fertilization, and examine the embryo's development to blastocyst (BL) stages have been made. The techniques even evaluate the implantation potential using static oocyte pictures throughout and after pregnancy [44].

14.6 Postpartum period

Another common gynecological disorder is pelvic floor dysfunction (PFD). Urine loss, pelvic organ prolapse, sexual dysfunction, and fecal incontinence are the main clinical symptoms. From this vantage point, a study investigated the use of ultrasound technology and rehabilitation instruction based on an AI algorithm in the recovery of postpartum pelvic organ prolapse. As a result, AI systems have positive effects while processing ultrasonic pictures. Patients with pelvic organ prolapse responded better to postpartum nursing after receiving pelvic floor rehabilitation therapy. Additionally, a variety of consumer-grade wearable gadgets, such as smart rings and watches, might capture semicontinuous physiological measurements like oxygen saturation, blood pressure, heart rate variability, normal heart rate, and body temperature. They also monitor other behavior indicators like sleep duration, quality, and position with other patients as well as activity levels. The monitoring of these physiological factors is beneficial for accurately diagnosing early pregnancy-related disorders such as gestational hypertension and preeclampsia. Heart rate monitoring, activity monitoring, and diagnostic procedures are the three areas where digital technology and AI are most often used during pregnancy. Incorporating specialized technologies would produce better results from conception to delivery. Using digital tools and ANN implementation, the better understanding of the metamorphosis of pregnancy (BUMP) research, for instance, attempts to get a full understanding of the patient's experiences with prepregnancy and pregnancy symptoms [45].

14.7 Conclusion and future perspective

The integration of ultrasonography and AI will greatly decrease the incidence of missing and incorrect diagnoses, and dramatically increase the standard of medical care to the benefit of patients. AI helps physicians diagnose a range of disorders and diseases. Although there have been notable advances in the use of AI in obstetrics and gynecology, the universality and efficacy of many models need to be further investigated. To solve the problem of restricted accuracy, many techniques have been studied, such as constructing ensemble algorithms, employing ultrasound films or alternative imaging modalities, using features from time-series data as a validation set, and so on. Also, clinicians require information to be able to eliminate or standardize subjective bias to prevent misdiagnosis and to establish objective, fair, and universal generalization standards due to the ongoing optimization and change of algorithms [46].

Contrarily, obstetric ultrasonography techniques related to AI are starting to be used in social work and education. For instance, a fetal ultrasound telemedicine service can connect a remote obstetric unit with a specialized fetal medicine clinic, allowing for the provision of high-quality ultrasound diagnostic and specialist consultation as well as a significant reduction in family costs and travel time [47]. This method has also shown to be helpful in cross-border consulting. Since there are still many poor nations and rural places where obstetric ultrasound is not available, telemedicine/telediagnostic services can expand access to obstetric ultrasound diagnostics in low-resource settings. It had been demonstrated to be in very good agreement with regular ultrasound practice. VR is emerging as a new method of simulation-based ultrasound training in obstetrics and gynecology that significantly enhances learning effectiveness and knowledge retention in fetal ultrasound teaching such as identifying fetal brain anomalies on ultrasound imaging or researching fetal development [48]. We think that AI will be able to contribute significantly more to the field of obstetric ultrasound with the advancement of techniques and multidisciplinary integration.

References

[1] P.M. Iftikhar, M.V. Kuijpers, A. Khayyat, A. Iftikhar, M. DeGouvia De Sa, Artificial intelligence: a new paradigm in obstetrics and gynecology research and clinical practice, Cureus (2020).

[2] G.S. Desai, Artificial intelligence: the future of obstetrics and gynecology, J. Obstet. Gynecol. India 68 (4) (2018) 326–327.

[3] F. Bellussi, T. Ghi, A. Youssef, G. Salsi, F. Giorgetta, D. Parma, G. Simonazzi, G. Pilu, The use of intrapartum ultrasound to diagnose malpositions and cephalic malpresentations, Am. J. Obstet. Gynecol. 217 (6) (2017) 633–641.

[4] M.A. Makary, M. Daniel, Medical error—the third leading cause of death in the US, BMJ (2016) i2139.

[5] R. Abinader, S.L. Warsof, Benefits and pitfalls of ultrasound in obstetrics and gynaecology, Obstet. Gynecol. Clin. N. Am. 46 (2) (2019) 367–378.

[6] C.L. Ondeck, D. Pretorius, J. McCaulley, M. Kinori, T. Maloney, A. Hull, S.L. Robbins, Ultrasonographic prenatal imaging of fetal ocular and orbital abnormalities, Surv. Ophthalmol. 63 (6) (2018) 745–753.

[7] Z. Chen, Z. Liu, M. Du, Z. Wang, Artificial intelligence in obstetric ultrasound: an update and future applications, Front. Med. 8 (2021).

[8] E.A. Krupinski, H.L. Kundel, SPIE medical imaging 50th anniversary: history of the Image Perception, Observer Performance, and Technology Assessment Conference, J. Med. Imaging 9 (S1) (2022), https://doi.org/10.1117/1.jmi.9.s1.012202.

[9] F. Dhombres, P. Maurice, L. Guilbaud, L. Franchinard, B. Dias, J. Charlet, E. Blondiaux, B. Khoshnood, D. Jurkovic, E. Jauniaux, J.-M. Jouannic, A novel intelligent scan assistant system for early pregnancy diagnosis by ultrasound: clinical decision support system evaluation study, J. Med. Internet Res. 21 (7) (2019).

[10] P.L. Hedley, C.M. Hagen, C. Wilstrup, M. Christiansen, The use of artificial intelligence and machine learning methods in early pregnancy pre-eclampsia screening: a systematic review protocol, PLoS One 18 (4) (2023) e0272465, https://doi.org/10.1371/journal.pone.0272465.

[11] X. Yang, L. Yu, S. Li, H. Wen, D. Luo, C. Bian, J. Qin, D. Ni, P.-A. Heng, Towards automated semantic segmentation in prenatal volumetric ultrasound, IEEE Trans. Med. Imaging 38 (1) (2019) 180–193.

[12] H. Ryou, M. Yaqub, A. Cavallaro, A.T. Papageorghiou, J. Alison Noble, Automated 3D ultrasound image analysis for first trimester assessment of fetal health, Phys. Med. Biol. 64 (18) (2019) 185010.

[13] S. Nie, J. Yu, P. Chen, Y. Wang, J.Q. Zhang, Automatic detection of standard sagittal plane in the first trimester of pregnancy using 3-D ultrasound data, Ultrasound Med. Biol. 43 (1) (2017) 286–300.

[14] M.C. Thomas, S.P. Arjunan, R. Viswanathan, Nuchal translucency thickness measurement in fetal ultrasound images to analyze down syndrome, IETE J. Res. (2021) 1–11, https://doi.org/10.1080/03772063.2021.1972847.

[15] M.D. Abràmoff, P.T. Lavin, M. Birch, N. Shah, J.C. Folk, Pivotal trial of an autonomous AI-based diagnostic system for detection of diabetic retinopathy in primary care offices, NPJ Digit. Med. 1 (1) (2018).

[16] P. Wang, T.M. Berzin, J.R. Glissen Brown, S. Bharadwaj, A. Becq, X. Xiao, P. Liu, L. Li, Y. Song, D. Zhang, Y. Li, G. Xu, M. Tu, X. Liu, Real-time automatic detection system increases colonoscopic polyp and adenoma detection rates: a prospective randomised controlled study, Gut 68 (10) (2019) 1813–1819.

[17] L. Drukker, R. Droste, P. Chatelain, J.A. Noble, A.T. Papageorghiou, Expected-value bias in routine third-trimester growth scans, Ultrasound Obstet. Gynecol. 55 (3) (2020) 375–382.

[18] G. Sciortino, D. Tegolo, C. Valenti, Automatic detection and measurement of nuchal translucency, Comput. Biol. Med. 82 (2017) 12–20, https://doi.org/10.1016/j.compbiomed.2017.01.008.

[19] Pluym, et al., Accuracy of automated three-dimensional ultrasound imaging technique for fetal head biometry, Ultrasound Obstet. Gynecol. 57 (5) (2021) 798–803, https://doi.org/10.1002/uog.22171.

[20] Yang, et al., Hybrid attention for automatic segmentation of whole fetal head in prenatal ultrasound volumes, Comput. Methods Prog. Biomed. 194 (2020) 105519, https://doi.org/10.1016/j.cmpb.2020.105519.

[21] R. Aggarwal, V. Sounderajah, G. Martin, D.S. Ting, A. Karthikesalingam, D. King, H. Ashrafian, A. Darzi, Diagnostic accuracy of deep learning in medical imaging: a systematic review and meta-analysis, NPJ Digit. Med. 4 (1) (2021).

[22] J. Torrents-Barrena, G. Piella, N. Masoller, E. Gratacós, E. Eixarch, M. Ceresa, M.Á. Ballester, Segmentation and classification in MRI and US fetal imaging: recent trends and future prospects, Med. Image Anal. 51 (2019) 61–88.

[23] J. Matthew, E. Skelton, T.G. Day, V.A. Zimmer, A. Gomez, G. Wheeler, N. Toussaint, T. Liu, S. Budd, K. Lloyd, R. Wright, S. Deng, N. Ghavami, M. Sinclair, Q. Meng, B. Kainz, J.A.

Schnabel, D. Rueckert, R. Razavi, J. Simpson, J. Hajnal, Exploring a new paradigm for the fetal anomaly ultrasound scan: artificial intelligence in real time, Prenat. Diagn. 42 (1) (2021) 49–59.

[24] S. Płotka, T. Włodarczyk, A. Klasa, M. Lipa, A. Sitek, T. Trzciński, FetalNet: Multi-task Deep Learning Framework for fetal ultrasound biometric measurements, Commun. Comput. Inf. Sci. (2021) 257–265.

[25] J. Yin, J. Li, Q. Huang, Y. Cao, X. Duan, B. Lu, X. Deng, Q. Li, J. Chen, Ultrasonographic segmentation of fetal lung with deep learning, J. Biosci. Med. 09 (01) (2021) 146–153.

[26] M.A. Maraci, C.P. Bridge, R. Napolitano, A. Papageorghiou, J.A. Noble, A framework for analysis of linear ultrasound videos to detect fetal presentation and heartbeat, Med. Image Anal. 37 (2017) 22–36.

[27] X.P. Burgos-Artizzu, D. Coronado-Gutiérrez, B. Valenzuela-Alcaraz, E. Bonet-Carne, E. Eixarch, F. Crispi, E. Gratacós, Evaluation of deep convolutional neural networks for automatic classification of common maternal fetal ultrasound planes, Sci. Rep. 10 (1) (2020).

[28] N. Baños, et al., Quantitative analysis of the cervical texture by ultrasound and correlation with gestational age, Fetal Diagn. Ther. 41 (4) (2016) 265–272, https://doi.org/10.1159/000448475.

[29] N. Baños, et al., Quantitative analysis of cervical texture by ultrasound in mid-pregnancy and association with spontaneous preterm birth, Ultrasound Obstet. Gynecol. 51 (5) (2018) 637–643, https://doi.org/10.1002/uog.17525.

[30] T.L.A. van den Heuvel, H. Petros, S. Santini, C.L. de Korte, B. van Ginneken, Automated fetal head detection and circumference estimation from free-hand ultrasound sweeps using deep learning in resource-limited countries, Ultrasound Med. Biol. 45 (3) (2019) 773–785.

[31] R.K. Morris, C.H. Meller, J. Tamblyn, G.M. Malin, R.D. Riley, M.D. Kilby, S.C. Robson, K.S. Khan, Association and prediction of amniotic fluid measurements for adverse pregnancy outcome: systematic review and meta-analysis, BJOG Int. J. Obstet. Gynaecol. 121 (6) (2014) 686–699.

[32] C.H. Jansen, C.E. Kleinrouweler, A.W. Kastelein, L. Ruiter, E. van Leeuwen, B.W. Mol, E. Pajkrt, Follow-up ultrasound in second-trimester low-positioned anterior and posterior placentae: Prospective cohort study, Ultrasound Obstet. Gynecol. 56 (5) (2020) 725–731.

[33] L. Bricker, A.T. Papageorghiou, B. Kemp, W. Stones, E.O. Ohuma, S.H. Kennedy, M. Purwar, L.J. Salomon, D.G. Altman, J.A. Noble, E. Bertino, M.G. Gravett, R. Pang, L.C. Ismail, F.C. Barros, Re: ultrasound-based gestational-age estimation in late pregnancy, Ultrasound Obstet. Gynecol. 48 (6) (2016) 693.

[34] A.T. Papageorghiou, B. Kemp, W. Stones, E.O. Ohuma, S.H. Kennedy, et al., Ultrasound-based gestational-age estimation in late pregnancy, Ultrasound Obstet. Gynecol. 48 (2016) 719–726.

[35] R. Saxena, G. Saxena, K. Joshi, S. Sunipa, K. Yadav, Foetal kidney length as a parameter for determination of gestational age in pregnancy by ultrasonography, Int. J. Med. Res. Prof. 2 (6) (2016), https://doi.org/10.21276/ijmrp.2016.2.6.013.

[36] Y. Jeong, S. Lee, D. Park, K. Park, Accurate age estimation using multi-task Siamese network-based deep metric learning for front face images, Symmetry 10 (9) (2018) 385, https://doi.org/10.3390/sym10090385.

[37] A.I.L. Namburete, R.V. Stebbing, B. Kemp, M. Yaqub, A.T. Papageorghiou, J. Alison Noble, Learning-based prediction of gestational age from ultrasound images of the fetal brain, Med. Image Anal. 21 (1) (2015) 72–86, https://doi.org/10.1016/j.media.2014.12.006.

[38] L. Bricker, A.T. Papageorghiou, B. Kemp, W. Stones, E.O. Ohuma, S.H. Kennedy, M. Purwar, L.J. Salomon, D.G. Altman, J.A. Noble, E. Bertino, M.G. Gravett, R. Pang, L.C. Ismail, F.C. Barros, Re: ultrasound-based gestational-age estimation in late pregnancy, Ultrasound Obstet. Gynecol. 48 (6) (2016) 693, https://doi.org/10.1002/uog.17355.

[39] A.T. Papageorghiou, et al., Ultrasound-based gestational-age estimation in late pregnancy, Ultrasound Obstet. Gynecol. 48 (6) (2016) 719–726, https://doi.org/10.1002/uog.15894.

[40] L.H. Lee, et al., Machine learning for accurate estimation of fetal gestational age based on ultrasound images, NPJ Digit. Med. 6 (1) (2023), https://doi.org/10.1038/s41746-023-00774-2.

[41] Á.S. Grétarsdóttir, T. Aspelund, Þ. Steingrímsdóttir, R.I. Bjarnadóttir, K. Einarsdóttir, Preterm births in Iceland 1997-2016: preterm birth rates by gestational age groups and type of preterm birth, Birth 47 (1) (2019) 105–114, https://doi.org/10.1111/birt.12467.

[42] S. Allen, C. Belcher, J. Newnham, The preterm birth prevention initiative - safely lowering the rate of preterm birth in Western Australia, Women Birth 31 (2018) S53, https://doi.org/10.1016/j.wombi.2018.08.158.

[43] N. Zaninovic, Z. Rosenwaks, Artificial intelligence in human in vitro fertilization and embryology, Fertil. Steril. 114 (5) (2020) 914–920, https://doi.org/10.1016/j.fertnstert.2020.09.157.

[44] C. Wakefield, M. Frasch, Predicting patients requiring treatment for depression in the postpartum period from common electronic medical record data antepartum using machine learning [ID: 1365796], Obstet. Gynecol. 141 (5S) (2023) 63S, https://doi.org/10.1097/01.aog.0000930572.02566.ab.

[45] R. Aggarwal, V. Sounderajah, G. Martin, D.S. Ting, A. Karthikesalingam, D. King, H. Ashrafian, A. Darzi, Diagnostic accuracy of deep learning in medical imaging: a systematic review and meta-analysis, NPJ Digit. Med. 4 (1) (2021).

[46] O. Russakovsky, J. Deng, H. Su, J. Krause, S. Satheesh, S. Ma, Z. Huang, A. Karpathy, A. Khosla, M. Bernstein, A.C. Berg, L. Fei-Fei, Imagenet Large Scale Visual Recognition Challenge, Int. J. Comput. Vis. 115 (3) (2015) 211–252.

[47] Z. Sobhaninia, S. Rafiei, A. Emami, N. Karimi, K. Najarian, S. Samavi, S.M. Reza Soroushmehr, Fetal ultrasound image segmentation for measuring biometric parameters using multi-task deep learning, in: 2019 41st Annual International Conference of the IEEE Engineering in Medicine and Biology Society (EMBC), 2019.

[48] S. Jain, Role of three dimensional ultrasound in evaluation of Foetal growth restriction, Clin. Radiol. Imaging J. 1 (1) (2017), https://doi.org/10.23880/crij-16000105.

Chapter 15

Deadly cancer of cervix tackled with early diagnosis using machine learning

Durairaj Mohanapriya[a], Kunnathur Murugesan Sakthivel[b], Nagendiran Baskar[b], H. Jude Immaculate[c], and Mariappan Selvarathi[c]

[a]*Department of Computer Science, PSG College of Arts & Science, Coimbatore, Tamilnadu, India,* [b]*Department of Biochemistry, PSG College of Arts & Science, Coimbatore, Tamilnadu, India,* [c]*Department of Mathematics, Karunya Institute of Technology and Sciences, Coimbatore, Tamilnadu, India*

15.1 Introduction

Cervical cancer is the most common disease among women after breast cancer and poses a major threat to both the physical and mental health of those affected. This malignant disease can be easily prevented if tested and diagnosed early. Although scientific developments have considerably improved clinical recognition of cervical cancer, precise diagnosis is still challenging. Cervical cancer identification and examination have advanced significantly due to the emergence of artificial intelligence (AI)-based clinical imaging techniques in recent years. The strengths of these methods include less time spent on tasks, a need for fewer specialists and technologists, and a lack of prejudice resulting from subjective elements. To improve the accurateness of early recognition, we used eXtreme gradient boosting (XGBoost), random forest (RF), decision tree (DT), support vector machine (SVM), K-nearest neighbor (KNN), naive Bayes (NB), multiple perceptron (MP), and logistic regression (LR) classifiers to determine if a patient has cancer. To perform a comparison analysis, we used data splitting, numerous metrics, statistical tests, and 10-fold cross validation. Evaluation measures show improvements in the total number of characteristics are reduced by using algorithms such as SVM, XGBoost, RF, DT, KNN, LR, NB, MP, J48 Tress, and LR. Furthermore, in this chapter we also discuss the opportunities and difficulties of applying AI for cervical cancer identification and examination.

Artificial Intelligence and Machine Learning for Women's Health Issues
https://doi.org/10.1016/B978-0-443-21889-7.00003-8

15.2 Statistics—Global and Indian scenario

There is a perception that cancer incidence is rising in India, and there is optimism that perhaps technological advancements will lead to more frequent cancer diagnoses, a shift in our mindset towards the disease, the dispelling of cancer-related myths, and an increase in our willingness to accept cancer diagnoses and have open conversations about it. The statistics show the epidemiology and cancer prevention efforts in world wide with a population of 1.36 billion [1]. Between the ages of 20 and 29, there are 137.8 million women and 230.5 million women, respectively. Whereas the phase-homogenous occurrence proportion of cervical cancer is 18.0 per 100,000 and the crude incidence proportion is 18.7 per 100,000, the prevalence of human papilloma virus (HPV) varies from 2.3% to 36.9%. According to estimates, there will be 9.6 million cancer deaths (9.5 million excluding nonmelanoma skin cancer) and 18.1 million new cancer cases (17.0 million excluding nonmelanoma skin cancer) in 2018. Lung cancer accounts for both the highest percentage of cancer cases diagnosed (11.6% of all cases) and the highest percentage of cancer deaths (18.4% of all cancer deaths). In terms of incidence, female breast cancer (11.6%), prostate cancer (7.1%), and colorectal cancer (6.1%) are closely followed by colorectal cancer (9.2%), stomach cancer (8.2%), and liver cancer (8.2%) in terms of mortality [2]. The country has a comprehensive national cancer control program that also has a separate noncommunicable disease program. There was a national cervical cancer initiative that ran from 2017 to 2022. Visual inspection with acetic acid (VIA) is an opportunistic screening technique that is advised. Women between the ages of 35 and 55 years are the target demographic for the project [3]. With 8.2 million cancer deaths and 32.6 million new cases of cancer among those older than 15 years over a 5-year period, the International Agency for Research on Cancer (IARC) reported that there were 14.1 million new cases of cancer globally in 2012, as expected (up from 12.7 million in 2008) [4]. By 2025, there will be a significant increase in new cancer cases, according to predictions from GLOBOCAN 2012. It is not unexpected that India followed a pattern like the rest of the developing world in terms of the number of new cases of cancer (57% or 8 million), fatal cancer cases (65% or 5.3 million), and the frequency of cancer cases over five years (48% or 15.6 million).

15.3 Current method available for cervical cancer diagnosis

According to the most recent World Health Organization (WHO) recommendations, HPV testing and pictorial scrutiny with acetic acid (PSA) are two techniques that should be used to detect cervical cancer early. Because PSA is only employed when the first two techniques are not available, we concentrated on the first two methods. Cervical cells that have been gently exfoliated are used as test samples for HPV testing and diagnostics. HPV testing finds high-risk varieties of HPV in the cervix, whereas morphologic inspection uses a magnifying glass to find cells taken from the cervix for probable cervical cancer or precancerous lesions [5]. To determine if the issue is cancer related, a medical expert will ask for more testing in case of displayed indications or diagnostic test results that increase the probability of cervical cancer. Typically, a medical evaluation that includes a pelvic and rectovaginal assessment will come first, followed by questions about the patient's personal and family medical history [6]. The medical professional may suggest diagnosis and treatment to identify whether the patient has been given a cervical cancer diagnosis and, if so, whether it has spread to other areas of the patient's body. The conclusions of these tests will also be beneficial for the doctor in establishing a treatment plan.

Cervical cancer can arise because of persistent high-risk HPV, and thus, HPV testing can help screen high-risk populations and find HPV infection. HPV sequencing will make it easier to identify the risk of females with favorable cervical affront outcomes and HPV DNA-optimistic outcomes, increasing accessibility to cervical cancer screening and treatment [7]. To increase precision and expand the use of HPV testing in cervical cancer prediction, artificial intelligence (AI) learning technology is being used. This study identified AI methods for detecting cervical cancer [8]. To evaluate the indicative inefficiency of AI procedures in ascertaining illnesses in medical imaging, deep learning (DL) has demonstrated a major influence on medical tomography. The following are the goals of this investigation:

- Making cervical cancer diagnoses more accurate by using machine learning (ML) algorithms to assess and categorize the condition.
- To identify the links between the elements furthermost prospective to result in cervical cancer.
- The aim of the survey is to gather information about women's worries regarding cervical cancer and share it with both readers and respondents.

15.4   Resolutions using machine intelligence to detect cervical cancer

To create an accurate screening model, the dataset must be of high quality. A categorization algorithm's performance can be hindered by missing entries in the data and biased evaluations, which can result in incorrect results [9]. As a result, we examined our dataset extensively to identify any lacking entries. We outline the subsequent in the sections that follow.

15.4.1   Dataset portrayal

For our investigation, we used a dataset from the UCI depository. The dataset included medical histories, routines, sexual history, and demographic data of 858 patients. It contained four target parameters (biopsy, cytology, Schiller, and Hinselmann) and 32 risk factors. Some people opted not to respond to some questions out of concern for their privacy and, as a result, the data has numerous missing entries [10]. The patients underwent four separate tests to identify cervical cancer. Biopsy is one of these tests and is used in this investigation because it is regarded as the gold standard [11]. The dataset is seriously skewed, with only 55 cases positive for cervical cancer compared to remaining 803 negative cases.

15.4.2 Proposed approaches

The caliber of the database upon which decision support systems are constructed has a substantial impact on their effectiveness. Missing values, duplicated characteristics, and unequal class distribution are frequently found in medical datasets. Before using data mining techniques, the data must be comprehensive in order to create a trustworthy predictive model.

Owing to the extreme discrepancy in the dataset, the classifier is fully swamped with occurrences belonging to the mainstream session and ignores examples belonging to the marginal session, classifying them as noise. In such unlikely events, classifiers could be unable to accurately recognize some occurrences of the marginal session, leading to a low sensitivity score. Yet, the minority class typically corresponds to the sick individuals in medical datasets [12]. Properly identifying these situations is vital since diagnosing an unwell individual as healthy might result in serious repercussions. It is of the utmost importance to obtain a high sensitivity score and a low false-negative rate when diagnosing a disease. A performance in prediction is evenly distributed between modules in the mainstream and the marginal. To move the bias away from the mainstream session, specific tactics must be used to create balance in the dataset. Researchers are looking for methods to address this class imbalance issue.

A common solution to the imbalanced classification problem is data-level modification. The idea for statistics-level adjustment is to somehow restructure the dataset to yield a further evenly distributed class dissemination. This can be done by either undersampling the occurrences of the mainstream session or oversampling the instances of the marginal session. This procedure is carried out in the data preprocessing stage and it denotes fundamental classifier [13]. Afterwards, any ML technique can be employed in conjunction with the resampled data. The balanced resampled dataset serves as the classifier's training ground, enabling it to anticipate outcomes similarly to conventional classification techniques. This enhances the classifier's overall prediction performance. However, the excellence of the occurrences retained after using the sampler strategies affects classification performance more so than dataset balancing.

Practitioners of medicine recognize cervical cancer as a theoretically incurable illness. Affected patients are put in jeopardy by late diagnosis and treatment. Due to social norms, the high cost of therapies, the lack of easily obtainable medical facilities, and the delayed onset of symptoms, suggested examination for illness recognition is difficult in both developed and developing countries [14]. Accurate intervention of numerous dissimilar infections, including cervical cancer, using ML is economical, computationally cheap, and effective. Patients are not obligated to undergo tedious, modern therapies, and AI can aid in the prompt identification of cervical cancer. The issue in question with currently available ML techniques for an illness diagnosis is how much they depend on the forecasting capacity of just a single classifier. Using just one categorization approach does not ensure the best prognosis because of bias, inappropriate fitting, insufficient handling of information that is noisy, and abnormalities [15]. This research proposes a categorization strategy based on mainstream voting to deliver a precise evaluation that addresses patient symptoms or issues.

A hybrid sampling technique is presented to solve the significant class imbalance found in the data [16]. For further performance improvement, a genetic algorithm (GA)-based feature variety technique is added into the process. The most essential properties were found using GA, which was then applied to simulation process. This study makes use of nine distinct classification algorithms: SVM, XGBoost, RF, DT, KNN, NB, MP, and LR classifiers are used to assess how well these classifiers work. Final results are verified in terms of accurateness, sensitivities, specificities, and z-score.

15.5 Classification methods/algorithms to analyze datasets for cervical cancer

The data must be of outstanding quality to create a reliable diagnostic model. A classifying algorithm's prediction performance can be decreased by the existence of missing items in the data, which can also provide skewed estimations and erroneous conclusions [17]. As a result, the type of data used in our study was meticulously analyzed to identify any missing entries. Next, using the cleaned-up dataset, we carried out the ensuing steps. We used a proposed hybrid sample selection strategy to address the considerable imbalance problem found in the data [18]. For additional performance improvement, we added an evolutionary algorithm-based attribute selection to the process. The most crucial properties were found using algorithms, which were then used to construct the system [19]. A wide range of classifiers, including DT, SVM, RF, KNN, NB, MP, and LR were extensively investigated. In comparison to single classification methods tested on the same benchmark datasets, the research findings indicate a precision rate of 94%, which is substantially higher. As a result, this suggested paradigm gives healthcare professionals the opportunity to obtain second opinions to help with early identification of disease as well as prompt treatment.

15.5.1 Process flow

Using a reference set of data for an illness that we were able to gather from the ML algorithms, we established an intelligent technique for predicting the possibility of cervical cancer [20]. First, we thoroughly examined the dataset for any missing values and eliminated samples that have numerous properties lacking. We handled the remaining missing entries using a multivariate imputation technique. We addressed the dataset's substantial class imbalance using an unfamiliar composite sampling plan of action. We used ML techniques used to identify the most important characteristics for the prediction task [21]. We were able to develop a trustworthy decision support system for the diagnosis of cervical cancer by combining all these methods. To further enhance the effectiveness of the classification methods, we employed ML algorithms to identify the key risk factors for cervical cancer. Hence, ML, like SVM, XGBoost, RF, DT, KNN, MP, NB, and LR algorithms, finds the key features to maximize class separation. Our suggested hybrid resampling technique efficiently tackles the imbalance problem and exhibits great efficiency for the diagnosis of cervical cancer (Fig. 15.1).

15.5.2 Support vector machine (SVM)

The appropriate characteristic selection for supervised classification, given the assortment of structures in a cataloging issue, is a fundamental issue. A few characteristic subgroup selection techniques have already been presented, even though several have been used in contexts with more than 100 dimensions. Our task's large feature dimension and sample complexity make computing of them impossible. As an alternative, we provide a based feature percentage of patients by extracting relevant metrics from the SVM's decision boundary. These indicators are used to rank the features, and a selection is then chosen using a statistically significant correlation test.

A problem's SVM decision function may be articulated as,

$$h(x) = w \cdot \Phi(x) + b = \sum_{i=1}^{n} \alpha_i \, y_i \, K(x, x_i) + b \qquad (15.1)$$

where $y_i \in R_d \, \{\pm 1\}$ is the session tag for X_i and $X_i \in R_d$ is the training exemplar. A transformation Φ (.) translates the participation space R_d data points x into the higher-dimensional feature space $R_d, (D \geq d)$ [22]. A kernel function $K(\cdot, \cdot)$ that specifies an internal invention in RD performs the mapping. By locating the higher dimensional space in the training dataset that is farthest to the adjacent photo (X_i) from the test dataset, which simplifies to resolution a linearly restricted convergent polynomial program, the parameters $\Phi(X_i)$ are optimized. SVM produces a nonlinear boundary $h(x) = 0$ in the contribution for the general case of nonlinear mapping.

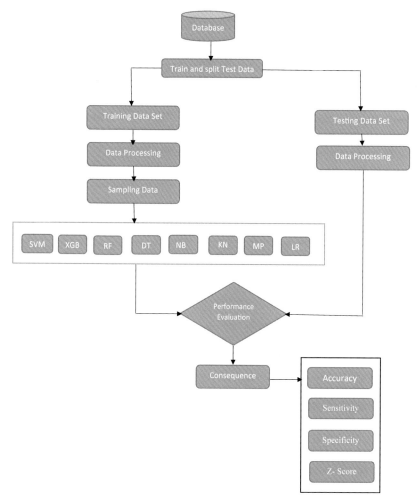

FIG. 15.1   Process flow.

15.5.3   eXtreme gradient boosting (XGBoost)

A tree-based aggregation tactic is known as XGBoost. Identification, prediction, and scoring issues may all be solved with XGBoost. Gradient boosting is a variant of XG boosting. It is a boosting form of a partnership called gradient boosting (GB), which creates predictors consecutively rather than independently. A strong classifier is created by merging weak classifiers using the GB approach. The goal of the GB is to create an iterative model that maximizes a gradient descent. With the use of variations in the loss function, it identifies the shortcomings of training examples.

$$y = ax + b + e \qquad (15.2)$$

where, the erroneous word is indicated by e. The loss function assesses the model's ability to accurately match the innovative data [23]. The error function depends on the optimization aim; for regression, it evaluates the discrepancy between the actual and predictor variables, whereas for categorization, it gauges how well a model categorizes instances. As a result of predictors learning from one another's prior errors, this method requires fewer repetitions and far less effort. To forecast values of the form, the GB instructs a model C. Ada boost is inferior to XGBoost in terms of speed and effectiveness. Compared to other conventional single ML techniques, it is 10 times quicker and extremely extensible. The patchy information is processed using XGBoost, which also employs several minimization and normalization strategies. Moreover, it makes use of the ideas of parallel and distributive computing.

15.5.4 Random forest (RF)

The RF may be thought of as an aggregation of DTs. By randomly selecting bootstrap samples from training sets and combining separately learned classifiers by a majority vote, the bagging technique reduces the variance of a model. RF is a tree-based collective learning methodology for regression as well as classification. Each tree is trained using a bootstrap specimen size, and from a randomly selected sample of all components, potentially the best solution factors are picked for each separation. The selection procedures are different for problems involving regression and classification. In the first instance, the Gini index was applied, and in the second, variance reduction.

$$\text{Gain } (T, X) = \text{Entropy } (X) - \text{Entropy } (T, X) \qquad (15.3)$$

Computing a plurality of votes or an aggregate has been used to estimate the multilingual prognosis of the RF for regression as well as classification. Regression analysis-like stochastic prognosis is possible if the extrapolation algorithm chooses to provide a binary outcome [24]. Formulation may be used to determine the mutual information for a RF, where T stands for the goal variable, X for the feature set that has to be divided, and Gain (T, X) for the values of the independent variable following the division of the feature set.

15.5.5 Decision tree (DT)

The nonsupervised learning method known as DT is frequently used for regression and classification issues. Using conventional decision-making processes and complex analytical characteristics, the goal is to increase the forecasting models of the posterior probability. An illustration of a marginal approximation is a tree. The sum of product (SOP) approach is used to depict it. An alternative term for SOP is disjunctive normal structure [25]. As a result, every branching out from a large tree root to a single subtree belonging to the same class is merely a juxtaposition of characteristics, and any divisions concluding in that class create a discontinuity. Formulation is a representation of an entropy E, where E stands for entropy, s for samples, Py for yes or no probability, Pn for no, and n for the number of samples in total.

$$E(S) = \sum_{k=0}^{n} \binom{n}{k} - Py * \log 2Pn \qquad (15.4)$$

15.5.6 K-nearest neighbor (KNN)

KNN is an example of a typical passive learner since it will remember the pattern rather than learning it. It will calculate how similar a pattern is to its K-nearest neighbors in terms of their distance, and it will categorize the patterns using frequencies that correspond to the neighbors' number. An approach for supervised learning that addresses classification issues is KNN. The important thing is to research each category's features beforehand [26]. The distance between the new person and all preceding people is taken into account by the KNN method, which uses the qualities taken from the categorization level to determine which k class is closest. The test data are then assigned to the KNN group, which includes the greatest number of individuals in a certain class. The optimal k number in the study is determined through experimentation, and distance is calculated using the Euclidean technique.

15.5.7 Multilayer perceptron (MLP)

With the aid of a feedforward algorithm, MLP is an artificial neural network (ANN) model that converts input sets into the proper output sets. The nodes that make up a multilayer perceptron (MLP) are arranged in layers, each of which is coupled to the layer before it. Except for input nodes, each node is a neuron or computational unit with an activation mechanism that is nonlinear [27]. The network of neurons undergoes training using the method of supervised learning, which is known as back propagation. The customized linear perceptron, or MLP, can provide information that cannot be separated by linearity.

15.5.8 Naive Bayes (NB)

Thomas Bayes, a British scientist, created the NB algorithm, a classification technique that makes use of probability and statistical techniques. Bayes' theorem describes how the NB algorithm forecasts opportunities for the future based on historical data. When the conditions between the qualities are thought to be independent, the theorem is joined with the Naive Bayes [28]. An extremely strong (naive) assumption of each condition/independence event is the fundamental characteristic of this NB classifier (NBC). The NBC is a specific kind of approximation, which is part of the wider and highly helpful Bayesian approximation. The prophecy accurateness of the NBC more sophisticated than that of other classifier models.

15.5.9 Logistic regression (LR) classifiers

The name "logistic regression" originated from the logistic equation, an equation that served as the system's fundamental element. The logistic technique, commonly referred to as the function of sigmoid, has been developed by statisticians for clarifying the ecological traits of population increases, which increase exponentially, and enhancing the sustainability of the environment. For much of the time, it resembles an S-shaped curve that is capable of transforming any number with an actual value to a quantity between 0 and 1, but never beyond those bounds [11]. A factor that predicts or set of predictors of a dichotomous variable that is dependent, such as the prevalence of cancer in patients, whether a person with cancer lived or died, or whether a patient reacted to therapy or not, can be assessed by employing LR to assess its consequence. Nominal, ordinal (ranked), ratio level (or continuous), interval, and other types of data are considered independent variables. The forecasting of the application of several predictors to a set of dependent variables is thus possible. Each self-care practice has a binary response option: "yes" or "no" (a "no" response indicates "missing" data). Using LR, models are built from the statistics that more accurately reflect the connections.

15.5.10 Result and analysis

In this chapter, we presented an innovative descriptive research approach for cervical cancer risk assessment. The cervical cancer risk depends on the level in the data gathered from the hospital. As already established, there is a substantial distinction between both favorable and adverse categorization in the collection of data at issue. The majority class is overwhelmingly biased in the attempt to identify cervical cancer from this unbalanced data, while the minority class has very poor prediction ability. Due to a substantial variance in both the sensitivity and the specificity values, this was apparent when the segregation experiment was carried out on the unsampled sample. When an attempt was made to categorize the dataset without sampling data. In Tables 15.1–15.4 we present the sensitivity, specificity, accuracy, and Z-score measurements we obtained for the unsampled data [29]. The sensitivity ratings for eight classifiers used in this investigation were all zero and the specificity values were approximately 100 [30].

TABLE 15.1 Z-scores (in percentages) for the unsampled data were obtained for sensitivity, specificity, and accuracy.

Algorithms	Sensitivity	Specificity	Accuracy	Z-score
SVM	0	99.71	92.96	0
XGBoost	11.45	98.14	92.03	33.52
RF	0	99.57	92.56	0
DT	9.64	96.57	90.44	30.51
KNN	0	99.28	92.3	0
NB	0	99.67	92.78	0
MP	0	96.23	91.7	0
LR	0	100	92.96	34.32

TABLE 15.2 Z-Score (in percentages) for the dispensary data were obtained for sensitivities, specificities, and accurateness.

Algorithms	Sensitivities	Specificities	Accurateness	Z-score
SVM	22.18	88.14	83.53	44.21
XGBoost	21.45	89.29	84.46	43.76
RF	7.45	96.57	90.31	26.82
DT	32.36	76.86	73.71	47.87
KNN	30.36	77.71	74.37	48.57
NB	31.23	87.23	82.11	49.23
MP	24.23	84.24	79.34	43.23
LR	38	77.43	74.63	54.24

TABLE 15.3 Z-scores (in percentages) for the hospital data were obtained for sensitivities, specificities, and accurateness.

Algorithms	Sensitivity	Specificity	Accuracy	Z-score
SVM	5.82	98.28	91.77	23.92
XGBoost	13.27	96.27	90.3	35.72

Continued

TABLE 15.3 Z-scores (in percentages) for the hospital data were obtained for sensitivities, specificities, and accurateness—cont'd

Algorithms	Sensitivity	Specificity	Accuracy	Z-score
RF	7.82	96.43	90.3	27.46
DT	19.64	93.86	88.58	42.93
KNN	2	96.86	90.17	13.92
NB	21.67	97.41	92.18	41.46
MP	12.56	93.45	87.32	28.34
LR	15.3	98.71	92.98	45.43

TABLE 15.4 Z-scores (in percentages) for the sampled data set from labs were obtained for sensitivities, specificities, and accurateness.

Algorithms	Sensitivities	Specificities	Accurateness	Z-score
SVM	26.18	90.43	85.92	48.66
XGBoost	24.91	89.43	84.87	47.20
RF	15.09	92.14	86.72	37.29
DT	40.18	82.43	79.42	57.55
KNN	47.46	72.29	70.52	58.57
NB	48.78	93.76	87.78	59.78
MP	23.67	71.23	69.34	46.34
LR	50.45	83.28	93.76	62.89

15.5.11 Accuracy

Accuracy is described as the total number of correctly associated predictions made using ML algorithms.

$$\text{Accuracy} = \frac{\text{True Positive(TP)} + \text{True Negative(TN)}}{\text{True Positive(TP)} + \text{False Positive(FP)} + \text{True Negative(TN)} + \text{False Negative(FN)}}$$

$$(15.5)$$

15.5.12 Sensitivity

Sensitivity is used for figuring out the proportion of tendencies that were anticipated to be beneficial but ultimately concluded. The projected outcome is as follows:

$$\text{Sensitivity} = \frac{(\text{TP})\text{True Positive}}{(\text{TP})\text{True Positive} + (\text{FN})\text{False Negative}} \quad (15.6)$$

15.5.13 Specificity

The ratio of accurately anticipated relationships between cervical cancer and other diseases is computed as shown in Figs. 15.2–15.5; Tables 15.1–15.4:

$$\text{Specificity} = \frac{(\text{TN})\text{True Negative}}{(\text{TN})\text{True Negative} + (\text{FP})\text{False Positive}} \quad (15.7)$$

15.5.14 Z-score

This score describes the similarity of cervical cancer by

$$Z - \text{score}(A, B) = \frac{d\,\text{short}\,(A, B) - \mu\,d\,\text{short}(A, B)}{\sigma d\,\text{short}(A, B)} \quad (15.8)$$

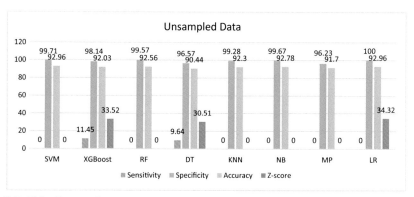

FIG. 15.2   Z-scores (in percentages) for the unsampled data were obtained for sensitivity, specificity, and accuracy.

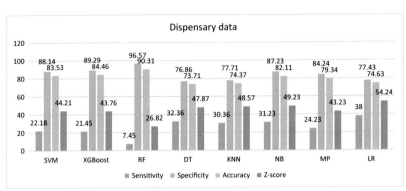

FIG. 15.3 Z-scores (in percentages) for the dispensary data were obtained for sensitivities, specificities, and accurateness.

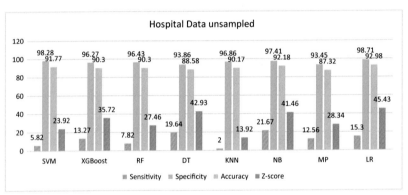

FIG. 15.4 Z-scores (in percentages) for the hospital data were obtained for sensitivities, specificities, and accurateness.

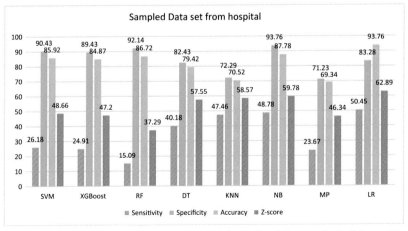

FIG. 15.5 Z-scores (in percentages) for the sampled dataset from labs were obtained for sensitivities, specificities, and accurateness.

15.6   Conclusion

In this chapter, we developed a sophisticated decision-making mechanism designed to help in early detection of cervical cancer. We concluded that the noticeable positive/negative class discrepancy in the collection of data used for the investigation could not be resolved by either sampling too much or inadequate sampling alone. In this study, we used cervical cancer data collected from a hospital to analyze and compare multiple ML algorithms to obtain the algorithms' efficiency using the measurement metrics of precision, recall, accuracy, specificity, sensitivity, and Z-score. Due to the practical significance of the proposed model, sensitivity and specificity have been mainly emphasized. Reliability is the proportion of instances that were correctly classified, whereas the two metrics of sensitivity and specificity measure the percentage of positive and negative examples that were correctly identified as such. The implementation of the work demonstrates that, in comparison to other conventional AI methods, the LR method has greater resilience and efficiency by applying Z-score. By employing deep learning algorithms, it is possible to expand the task to the following sentence by automatically identifying and categorizing features and classifying data with less attributes.

References

[1] M. Singh, et al., Secular trends in incidence and mortality of cervical cancer in India and its states, 1990-2019: data from the global burden of disease 2019 study, BMC Cancer 22 (1) (2022) 1–12.

[2] S. Kaur, et al., Callenges in cervical Cancer prevention: real-world scenario in India, South Asian J. Cancer 12 (01) (2023) 009–016.

[3] Web Source: https://prescriptec.org/countries/india/.

[4] National Health Commission Of The People's Republic Of China, National guidelines for diagnosis and treatment of cervical cancer 2022 in China (English version), Chin. J. Cancer Res. 34 (3) (2022) 256–269, https://doi.org/10.21147/j.issn.1000-9604.2022.03.06.

[5] X. Hou, et al., Artificial intelligence in cervical cancer screening and diagnosis, Front. Oncol. 12 (2022).

[6] B. Nithya, V. Ilango, Evaluation of machine learning based optimized feature selection approaches and classification methods for cervical cancer prediction, SN Appl. Sci. 1 (2019) 1–16.

[7] A. Mudawi, Naif, and Abdulwahab Alazeb., A model for predicting cervical cancer using machine learning algorithms, Sensors 22 (11) (2022) 4132.

[8] L. Akter, et al., Prediction of cervical cancer from behavior risk using machine learning techniques, SN Comput. Sci. Vol. 2 (2021) 1–10.

[9] F. Khanam, M.R.H. Mondal, Ensemble machine learning algorithms for the diagnosis of cervical Cancer, in: 2021 International Conference on Science & Contemporary Technologies (ICSCT), IEEE, 2021.

[10] A. Newaz, S. Muhtadi, F.S. Haq, An intelligent decision support system for the accurate diagnosis of cervical cancer, Knowl.-Based Syst. 245 (2022) 108634.

[11] N. Razali, et al., Risk factors of cervical cancer using classification in data mining, J. Phys. Conf. Ser. 1529 (2) (2020). IOP Publishing.

[12] J.W.-C. Hsu, et al., Classification of cervical biopsy free-text diagnoses through linear-classifier based natural language processing, J. Pathol. Inform. 13 (2022) 100123.

[13] W. William, et al., A review of image analysis and machine learning techniques for automated cervical cancer screening from pap-smear images, Comput. Methods Prog. Biomed. 164 (2018) 15–22.

[14] R.F. Josephine, E.V. Maani, C.J. Dunton, B.W. Jack, Cervical Cancer, 2022, StatPearls [Internet]. Last Update: November 2 https://www.ncbi.nlm.nih.gov/books/NBK431093/.

[15] Q.M. Ilyas, M. Ahmad, An enhanced ensemble diagnosis of cervical cancer: a pursuit of machine intelligence towards sustainable health, IEEE Access 9 (2021) 12374–12388.

[16] C. Zhao, et al., Improving cervical cancer classification with imbalanced datasets combining taming transformers with T2T-ViT, Multimed. Tools Appl. 81 (17) (2022) 24265–24300.

[17] J.J. Tanimu, et al., A machine learning method for classification of cervical cancer, Electronics 11 (3) (2022) 463.

[18] R. Gupta, A. Sarwar, V. Sharma, Screening of cervical cancer by artificial intelligence based analysis of digitized papanicolaou-smear images, Int. J. Contemp. Med. Res. 4 (5) (2017) 2454–7379.

[19] A. Shetty, V. Shah, Survey of cervical cancer prediction using machine learning: a comparative approach, in: 2018 9th International Conference on Computing, Communication and Networking Technologies (ICCCNT), IEEE, 2018.

[20] L. Zhang, et al., Intelligent diagnosis of cervical Cancer based on data mining algorithm, Comput. Math. Methods Med. 2021 (2021).

[21] Abisoye, O. Aderiike, et al., The prediction of cervical cancer occurence using genectic algorithm and support vector machine, in: 3rd International Engineering Conference (IEC 2019), 2019.

[22] I.U. Khan, et al., Cervical cancer diagnosis model using extreme gradient boosting and bioinspired firefly optimization, Sci. Program. 2021 (2021) 1–10.

[23] Y. Ling, et al., Application and Comparison of Several Machine Learning Methods in the Prognosis of Cervical Cancer, 2022.

[24] U.K. Lilhore, et al., Hybrid model for detection of cervical cancer using causal analysis and machine learning techniques, Comput. Math. Methods Med. 2022 (2022).

[25] S.R. Narayan, S. Vallee, R. Jemila Rose, Cervical Cancer detection based on novel decision tree approach, Comput. Syst. Sci. Eng. 44 (2) (2023) 1025–1038.

[26] P.K. Malli, S. Nandyal, Machine learning technique for detection of cervical cancer using k-NN and artificial neural network, International Journal of Emerging Trends & Technology in Computer Science (IJETTCS) 6 (4) (2017) 145–149.

[27] O.Y. Dweekat, S.S. Lam, Cervical cancer diagnosis using an integrated system of principal Component analysis, genetic algorithm, and multilayer perceptron, Healthcare 10 (10) (2022). MDPI.

[28] M. Girdonia, R. Garg, M.S. Pooja Jeyabashkharan, Minu., Cervical Cancer prediction using Naïve Bayes classification, Int. J. Eng. Adv. Technol. 8 (4) (2019) 2249–8958.

[29] M.D. Mohanapriya, R. Beena, Improving topic modelling for prediction of drug indication and side effects, Ann. Romanian Soc. Cell Biol. (2021) 11542–11558.

[30] D. Mohanapriya, D.R. Beena, Enhancing prediction of drug indication and side effects through named entity recognition and jointly learning of syntactic structures of sentences, Eur. J. Mol. Clin. Med. 7 (6) (2020) 170–176.

Chapter 16

AI, women's health care, and trust: Problems and prospects

Vaishali Singh
XIM University, Bhubaneswar, India

16.1 Introduction

Many industries, including the healthcare sector, have the potential to change as a result of artificial intelligence (AI) and machine learning (ML). The management of health care can benefit greatly from AI and ML, and this is especially true for the health care of women. To help with disease diagnosis and treatment, AI algorithms can initially analyze patient data, such as medical imaging, electronic health records (EHRs), and genetic data. There are myriad ways in which AI and ML can lead to effective healthcare for women. Firstly, deep learning (DL) algorithms, for instance, may examine medical pictures like X-rays and CT scans to find anomalies and help with the identification of diseases like cancer. Similar to this, ML algorithms can examine EHRs to find individuals who are at a greater risk of contracting specific ailments and assist physicians in making better treatment decisions by identifying these patients [1]. Secondly, AI and ML can help with medication discovery and the creation of individualized therapy regimens. They can find potential new drug candidates and forecast the effectiveness of current medications for certain patient populations by analyzing enormous quantities of biological and chemical data. Thirdly, AI and ML can provide remote patient monitoring, enabling healthcare professionals to keep an eye on patients away from conventional clinical settings. AI algorithms can analyze data from wearable sensors and gadgets to identify possible health hazards and take action before major difficulties arise. These data include symptoms, levels of activity, and other health indicators [2,3]. Fourthly, medical image analysis using AI and ML can spot trends and abnormalities that human observers might overlook. For instance, they can examine mammograms to see early breast cancer symptoms or retinal pictures to spot diabetic retinopathy symptoms. Finally, administrative duties like organizing appointments

**Artificial Intelligence *and* Machine Learning
for Women's Health Issues**
https://doi.org/10.1016/B978-0-443-21889-7.00002-6

235

and filing insurance claims may be automated by AI and ML, freeing up physicians' time to concentrate on patient care. AI may, for instance, assist in the early diagnosis of breast and cervical cancers, identify pregnancy difficulties, and forecast the likelihood of postpartum depression for women, all of which can enhance healthcare outcomes. Additionally, it can help with the creation of individualized treatment programs for female patients with long-term diseases including diabetes and heart disease [4,5].

While it is true that AI and ML continue to make enormous strides in scientific research, medical advancements, therapeutic upgrade, and curative innovation, the future scenario in terms of widespread acceptance and use of these evolving tech-enabled changes in health care seems like a far-off dream. The potential benefits of ML also include tradeoffs, which might be dangerous, especially for women, if not managed appropriately. For instance, overconfidence in the performance of an ML algorithm's conclusions in situations other than those that are evident in the data might result in significant financial losses [6]. More dangerously, poorly implemented and interpreted ML algorithms can sustain fundamental racial and ethnic inequalities across genders or even within the same gender grouping. Although women have a lot to gain from using AI and ML, there is still a gender bias in both their development and application. In the domains of AI and ML, women are disproportionately underrepresented. Only 21% of women worldwide work in AI and ML, according to the Hacker-Rank 2021 Women in Tech Report. According to a 2021 analysis by the AI Now Institute, women account for just 15% of AI research employees in the United States at Facebook and 10% at Google. Only 18% of data science and AI workers in the United Kingdom are women, according to a 2021 Deloitte survey. In a 2019 study by AI4ALL, a group that seeks to enhance diversity in AI, just 22% of those surveyed who declared themselves AI professionals were women [7,8]. So, the impact of AI and ML can vary significantly across men and women and even across women with differences in race, place of residence, education, and health needs.

From a gendered perspective and looking specifically at the healthcare industry, the biggest challenge being faced by the AI and ML technological frameworks is their inability to incorporate ethics and trust in the system. Technology needs to be secure and utilized responsibly in order to be trusted. Making fair, egalitarian, and risk-aware policy choices with major consequences for people requires access to reliable, open, and accountable information. To evaluate the findings of any AI enabled health study, it is essential to have knowledge of the data, the model and its performance, and any extra context (such as location, gender, race, ethnicity, and exposure to disease-prone environments). People's mistrust of new technology is a result of their lack of awareness of AI's ability to assess these contextual factors. Most importantly, a rise in mistrust of AI and ML technology is being attributed to issues with security, transparency and value-based ethics.

Considering the gender digital divide that already pervades the world of technology, the issues related to trust, faith, and confidence of women in the emerging and rapidly changing technological landscape, especially in the important field of health care, are of critical concern. In fact, lack of faith in the emerging technologies is a significant contributing factor to women's underuse of AI and ML. Women are more likely to believe that technology is out of control, citing the rapid speed of development, and women worry more about not being able to distinguish between actual and bogus information. To prevent issues like prejudice and sexism, the digital abilities gap, and technologically driven displacement from undermining the opportunities provided by AI and ML, technology makers and businesses that rely on technology need to do more so that the decline in women's trust in our institutions, particularly in business and government, is not exacerbated.

The trust of managers, patients, and professional end users is crucial for the success of digital health. It is anticipated to improve the efficiency and security of the medical system while lowering rising healthcare expenses. But, until now, there has not been a lot of research into what the pillars of confidence in digital health systems are. As with many other human cognitive functions, trust is complex and highly contextual, and we stand to gain tremendously from research into these trust traits by encouraging wider implementation of this new and emerging technology.

The growth of our diversity and inclusion initiatives will be hindered or even stopped if we do not incorporate a technology and trust perspective into them. Gender inclusion and diversity has been a main agenda of policymakers for a long time. Therefore, there is considerable potential for a gender diversity plan that takes technological challenges into account, such as enhanced consumer and employee trust, improved access to talent, and new and improved goods and services. This chapter fills the gap by looking into the trust barriers in AI and ML from a gendered perspective.

The remaining chapter is organized into six parts. The next section deals with the broad contours at the intersection between technology and trust in terms of theories and conceptual frameworks. The third section investigates the gender bias and gender divide in use of technology and in trusting technology. The fourth section looks at the role and prospects of AI and ML for improving women's health care. The fifth section deals with the trust issues related to the use of AI and ML, especially in the healthcare sector. The last section outlines the ways to mitigate trust barriers and probability of bias in AI and ML as applied in the healthcare sector, particularly for women.

16.2 Overview of technology and trust

Technology and trust are fundamentally based on the concept of human-automation interaction. The term "human-automation interaction" describes how people engage with automated systems, including robots, AI, and other types of automation. Depending on the technology and the purpose for which it is being used, this contact might take various shapes. Here are a few illustrations:

1. Collaboration: In some situations, automated systems and people coexist in a collaborative engagement. For instance, a robot may help a human employee in a factory by carrying out tedious or risky jobs while the human supervises and controls the robot's activities [9].
2. Autonomous contact: In other circumstances, an automated system may function without human direction or oversight. For instance, a chatbot powered by AI may engage with clients on a website, replying to their questions and helping them with appropriate guidance [10].
3. Automation may occasionally be utilized to augment human talents during contact. By using a robotic device, a physician may, for instance, execute a sensitive procedure like surgery with increased precision and accuracy [11,12].
4. Interaction control: Occasionally, people oversee managing and guiding automated systems. For instance, a drone pilot may direct a drone's flight path and camera while collecting data or carrying out a search and rescue mission [13].

The demands of the user and the individual situation must be considered during the meticulous design of any human-automation interface. Additionally, it calls for open lines of communication and comprehension between humans and robots, as well as confidence in and openness about the system's capabilities and limits. When human-automation contact is well designed, the results can be more successful, fruitful, and secure.

There are several theories of trust and technology that explain how trust is formed, maintained, and lost in the context of technology. The generally accepted idea that describes how people acquire and use technology is called the "technology acceptance model" (TAM). TAM claims that two important aspects—perceived utility and conceived ease of use—are the foundations of technological trust. If technology is perceived by users as beneficial and simple to use, they are more inclined to trust it and adopt it. [14] Another idea is the Social Exchange Theory (SET), which describes how trust develops in social connections, particularly those involving users and technology. According to SET, technology trust is built on social exchange, in which consumers provide their personal information to the system in exchange for advantages like efficiency or ease. Users develop trust when they believe the advantages of sharing their assets exceed the dangers or expenses [15]. In terms of perception of trust, there is the Cognitive-Affective Theory of Trust (CATT). This theory describes how trust develops and is preserved in interpersonal interactions. Technology trust is founded on cognitive and emotive elements, according to CATT. Affective elements include users' emotional responses to technology, such as emotions of comfort or worry, and cognitive variables include users' views about the competency and dependability of technology. When people have favorable cognitive and emotive experiences with technology, trust is established and sustained [16]. These ideas shed light on how trust develops and is preserved in the technological environment. Understanding these ideas will help technology suppliers and developers create goods and services that encourage user adoption and trust.

There are many dimensions of trust in technology, as shown in Fig. 16.1. There are primarily three sets of characteristics: human, environment, and technology. The influence of human traits and environmental features will be comparable regardless of who or what the trustee is and whether it is a human, a type of AI, or an object such as an organization or a virtual team. A person with a high level of trustworthiness might, for example, be more willing to embrace and rely on new tools or team members [17]. In a similar vein, it will be simpler for a technology offered by a reputable institution or organization to win consumers' confidence than it would be for a comparable technology from a less reputable institution or organization. Human traits primarily consider the personality, trustworthiness, and risk-taking propensity of the individual. The nature or inclination of the trustor, which varies depending on events, kind of person, and cultural contexts, can be viewed as the overall tendency to trust others. The term "ability" often refers to a trustee's competency or set of abilities needed to carry out activities in a certain subject. For example, if a staff member is skilled in contract negotiations, the management can put their faith in them. The focus of environmental characteristics is on variables like the tasks' nature, cultural context, and institutional aspects. There are many distinct types of tasks. They could be minor or extremely vital, for instance. Race, religion, social class, and ethnicity may all influence one's cultural heritage. According to the research, there are two basic components of institutional factors: situational normalcy and structural guarantees. Situational normalcy refers

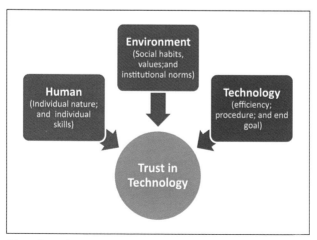

FIG. 16.1   Dimensions of trust in technology.

to the fact that everything is in order and the situation is normal. Contextual elements like promises, contracts, guarantees, and laws are referred to as structural assurances. Three viewpoints may be used to analyze a technology's characteristics: its performance, its process or traits, and its goal. Although human and environmental traits are generally comparable regardless of the trustee, the technological traits affecting trust for AI, ML, and robots will be different than they are for other items or humans.

Fig. 16.2 shows that in scholarly research on AI and trust, the most common factors responsible for influencing trust in human-AI interaction or collaboration [18] are complex algorithm, data sensitivity, cognitive bias, lack of operating knowledge, skepticism regarding future relevance/role of current technology, and uncertainty about the robustness of AI. Each factor is influencing the other factors in a loop, which makes the trust barrier a complex cycle to overcome. Health care is highly variable across different segments of society. AI can simulate human diagnostic care based on a particular set of data, and the processing of data through algorithms usually suffers from generalization bias especially for vulnerable groups such as women and minorities. Several human elements, including user education, past encounters, user prejudices, and opinions regarding automation, might affect users' trust in AI. While AI and ML score well in their computational capacity and precision, they still must navigate the trust barriers, especially in relation to the analytical and intuitive abilities of human healthcare experts.

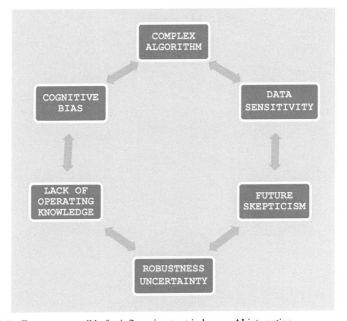

FIG. 16.2 Factors responsible for influencing trust in human-AI interaction.

16.3 Gender divide in trusting technology

Biases based on data and sociocultural prejudice are two of the most significant AI biases. As shown in Fig. 16.3, when an algorithm mines biased data, it is said to have data bias. When society's standards create cognitive biases in humans, this is known as social bias. As part of their profession, data scientists examine variables and presumptions in all datasets. Due to societal conventions, gender norms, or cultural norms, societal biases are factors that are sometimes neglected. These false presumptions may produce biased results and widen inequality. The development of new algorithmic technologies increasingly recognizes the importance of societal issues, yet "AI scientists continue to demonstrate a limited understanding of society" [19]. One of the key worries about the "rise of AI" has been a corresponding loss of human agency, and much of the discourse addressing algorithmic technology continues to be marked by a feeling of inevitability and technical determinism [20]. Three primary kinds of biases have been identified. Bias exists firstly, in the user's understanding of who they are and how they might use the software; secondly, in the data used to power the software, which makes inaccurate suggestions to the user; and thirdly, in the product's design, which makes it unattractive or difficult for certain kinds of users. Furthermore, the Internet of Things (IoT) and the "tracking and the datafication process of the body and routine behaviors such as running, resting, walking, and eating" are two examples of how gender inequalities have drawn attention.

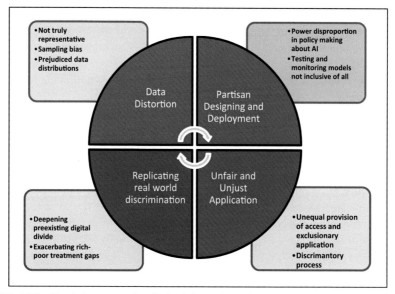

FIG. 16.3 Discriminatory practices in the design and use of AI in health care.

Evidence points to a potential gender gap in how much women and men trust technology. Women were less inclined than males to believe in machines with intelligence in a range of scenarios, including medical care, schooling, and transportation, according to a 2021 poll by Edelman. Women were shown to be less willing than men to trust driverless vehicles in a 2020 Morning Consult study. In contrast to 46% of men, just 32% of women stated they would feel comfortable in a self-driving automobile. Women are less inclined than males to trust social media businesses to secure their personal information, according to a YouGov poll from 2019 [21]. Compared to 22% of males, just 16% of women said they believed social media firms "a great deal" or "a lot." According to a 2018 Pew Research Center research, women are more likely than males to voice worries about how technology is affecting society [22]. For instance, as opposed to 43% of males, 53% of women expressed concern about large employment losses as a result of automation. These results imply that women may have greater skepticism than males regarding technology and its influence on society. However, it is crucial to remember that these gender disparities are not absolute and that a person's level of faith in technology can be influenced by a variety of circumstances [23].

The degree to which women perceive a risk associated with technology depends on a variety of elements, including culture, education, age, social level, and personal experiences. However, studies have revealed that women often approach technology with greater caution than men do [24]. Because they are worried about their privacy, security, and the possible negative repercussions on their personal and professional life, they could be more risk averse and slower to accept new technology. According to certain research, cultural norms and gender stereotypes may also have an impact on how risky women perceive technology to be. For instance, women's socialization to risk aversion and lower technological confidence may have an impact on how likely they are to test new technology [25]. Women may develop a false and unwarranted perception of inferiority in ICT if the area is portrayed as being mainly controlled by men. According to a review of the research [26] on how men and women utilize technology differently, women experience higher anxiety while using IT than men do. This anxiety lowers women's self-efficacy and raises views that using IT requires more work. Impostor syndrome, or a fear of failure, affects women in genuine ways, and men's mocking or dismissive responses to women's unease with technology make many women less likely to take part.

Healthcare device, gadget-based screening, electronic health care, and health information and communication technology firms are part of the expanding area of health technology, which is positioned directly between the high-technology and healthcare sectors. Comparatively, the high-technology sector still has a sizable gender gap—only around 25% of its employees are female. Although this industry is making efforts to improve its unfavorable gender diversity record, the proportion of women working in sophisticated technology has not increased significantly, although job possibilities are plentiful and salaries are often good [27].

According to E.J. Topol, one of the primary motivating constructs behind the usage of digital technologies is self-efficacy in using them [28]. The degrees of confidence in one's ability to learn and employ digital abilities seem to vary between men and women. In contrast to 67% of males, just 54% of women are favorable about robots and AI. In addition, women tend to be less knowledgeable than males about new technology, which may explain why they are more skeptical about them. In the context of AI, 41% of women compared to 53% of males had heard, read, or seen some knowledge concerning AI in the previous year. Other technological subjects likewise exhibit a gender imbalance.

Numerous study results [29] point to the exclusivity of digital technology design and a lack of experimentation on women as factors in women's decreased confidence in technology. For instance, in-depth research has looked at gender-based variations in the motion sickness brought on by exposure to virtual reality. Because virtual reality headsets were not created with female physiology in mind, a recent study showed that interpupillary distance led to motion sickness in women.

Fig. 16.4 shows the positive changes that can come if there is no gender bias in AI. Women possess the unique ability to bring about a collaborative leadership style. This is particularly important to build trust and create value for the user. Customers favor female AI over male AI because they perceive it to possess more positive human traits, such as warmth, compassion, and empathy [30].

16.4 Women's health care and AI: The prospects

Health care in general, and women's health care specifically, has the potential to be revolutionized by AI. As shown in Fig. 16.5, AI may be used to personalize treatment programs, increase the efficacy and accuracy of diagnosis, and even foresee prospective health problems [31,32]. The following sections discuss some applications of AI in women's health care.

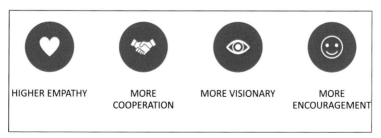

HIGHER EMPATHY   MORE COOPERATION   MORE VISIONARY   MORE ENCOURAGEMENT

FIG. 16.4 How more women AI teams can build trust for the healthcare system.

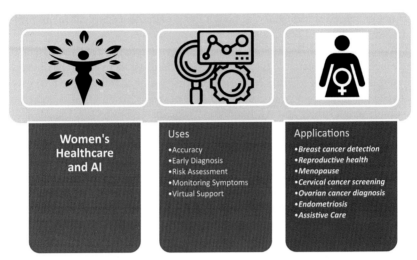

FIG. 16.5 Uses and applications of AI in women's health care.

Breast cancer detection: With an elevated level of accuracy, AI systems can analyze mammograms to identify breast cancer at its earliest stages. In doing so, women may benefit from immediate medical attention and have a greater chance of surviving. Here are several applications of AI and ML in the identification of breast cancer:

(a) Automated mammography analysis: AI systems may examine mammograms and spot anomalies like lumps or deposits that can be signs of breast cancer. This can facilitate quicker and more accurate breast cancer detection by radiologists.

(b) Computer-aided diagnosis: AI and ML algorithms can help radiologists in making breast cancer diagnoses by providing a second opinion. These algorithms can analyze mammograms and biopsy results and provide additional information that can help radiologists make more accurate diagnoses.

(c) Risk assessment: AI and ML models can evaluate patient information including age, family history, and lifestyle choices to estimate a woman's likelihood of breast cancer. This can assist medical professionals in creating unique screening and preventive plans for every patient.

(d) Recommendation for treatment: AI and ML systems can assist medical professionals in choosing the most effective course of action for patients with breast cancer. Based on variables including tumor size, stage, and subtype, these algorithms may examine patient data and suggest the best courses of action.

Reproductive health: To obtain understanding of women's periods and ovulation patterns, AI may be utilized to analyze data from wearables and fertility applications. Women can use these to monitor their fertility and increase their chances of getting pregnant. Here are some applications of AI and ML in reproductive health:

(a) Monitoring fertility: Women's period cycles and ovulation rhythms can be better understood by using wearable technology and AI and ML computations to track their fertility. Women can use this to monitor their fertility and increase their chances of getting pregnant.

(b) Predicting ovulation: Using information from fertility tracking equipment like basal body heat monitors or ovulation prediction kits, AI and ML algorithms can determine when a woman is probably going to ovulate. This may assist couples schedule their sexual activity to increase their chances of getting pregnant.

(c) Personalized fertility therapies: Personalized therapies are possible because AI and ML calculations can analyze patient information including weight, age, and medical history. This information can assist medical professionals in creating treatment programs that are personalized for each patient's requirements.

(d) Pregnancy monitoring: AI and ML algorithms can monitor fetal development during pregnancy and predict potential complications. This can help healthcare providers detect potential problems early and develop treatment plans to address them.

Conceiving and giving birth: AI can help in tracking fetal growth throughout pregnancy and foresee probable issues during childbirth. It may also be used to identify women who are at a high risk of premature labor by analyzing data from EHRs. Here are some applications of AI and ML in pregnancies and childbirth:

(a) Fetal monitoring: By examining ultrasound pictures, fetal heart rate sequences, and other data, AI and ML systems can keep track of the growth of the fetus during pregnancy. This can assist medical professionals in early detection of possible issues including fetal distress or growth limitation and the development of treatment strategies to address them.

(b) Preterm labor forecast: This involves identifying women who may be at a greater risk of preterm labor using AI and ML models that analyze information gathered from digital medical records. This can assist medical professionals in creating individualized treatment regimens that either avoid preterm birth or better manage it.

(c) Predicting gestational diabetes: A woman's likelihood of developing gestational diabetes during pregnancy may be predicted using AI and ML algorithms that analyze patient information including height and weight, age, and history of illness. This can assist medical professionals in creating individualized treatment regimens to control the disease and avoid problems.

(d) Planning a personalized birth: AI and ML systems may examine patient information including the mother's age, health history, and fetal development to provide personalized delivery schedules. This can assist healthcare professionals in creating strategies that are customized to each patient's unique needs, such as arranging a scheduled C-section for at-risk births.

Menopause: Women going through menopause can create individualized treatment programs using AI and ML. To suggest substitute hormones and other therapies, AI and ML can analyze information from digital medical records, genetic testing, and lifestyle variables. Here are some applications of AI and ML in menopause:

(a) Management of symptoms: AI and ML methods can assist medical professionals in creating individualized therapy programs to treat menopausal symptoms such hot flashes, mood swings, and sleep difficulties. The best suitable therapy for each patient can be suggested using these algorithms' analysis of patient information, including medical history, symptoms, and lifestyle variables.
(b) Forecasting the start of symptoms: AI and ML systems can analyze patient data to predict when women are expected to suffer menopausal symptoms. This can assist medical professionals in creating preventive treatment regimens that can lessen the intensity or persistence of symptoms.
(c) Health risk assessment: AI and ML algorithms can analyze patient data to estimate the likelihood that women will have menopause-related health problems including osteoporosis and cardiovascular illness. This can assist medical professionals in creating individualized preventative programs and recommending the right screening and testing.
(d) Assessment of quality of life: AI and ML models can examine patient data to evaluate the quality of life of women going through menopause. This can assist medical professionals in creating treatment programs that give quality of life enhancements and symptom control first priority.

Cervical cancer screening: AI and ML algorithms can examine cervical smear samples to find aberrant cells that could be signs of cervical cancer. These algorithms may be trained to spot minute variations in cell shape and locate patterns that the naked eye would miss.

Ovarian cancer diagnosis: A diagnosis of ovarian cancer may be made using AI and ML computations that examine patient information such as medical history, indications, and biomarkers. These algorithms may also be used to track the development of an illness and find the best available treatments.

Detect endometriosis using AI and ML: Endometriosis is a disorder in which the tissue that typically lines the lining of the uterus develops outside of it. Imaging data, medical history, and symptoms are all used to diagnose endometriosis. These algorithms may also be used to track the development of an illness and find the best available treatments.

Assistive care: AI and assistive devices can significantly enhance the care of women, especially in the areas of maternity health, female wellbeing, health, and chronic illnesses that are more common in women [32]. Here are some instances of how AI and assistive technology are being utilized to give women better care:

(a) Personalized care: Based on each woman's unique health requirements and preferences, personalized care for women may be delivered using AI and assistive technology. This can involve the use of remote monitoring tools that measure health indicators and notify healthcare practitioners of any problems, as well as online medical assistants that offer individualized advice and direction on health care [33–35].
(b) A more accurate diagnosis may be made for illnesses that afflict women by using AI and machine intelligence systems to analyze medical imaging and other diagnostic procedures. By analyzing mammograms and seeing subtle alterations in the breast tissue that may be symptomatic of cancer, for instance, AI algorithms can aid in the more accurate diagnosis of breast cancer.
(c) Management of chronic diseases: Conditions including coronary artery disease, diabetes, and autoimmune disorders that affect more women than men can be managed with the use of AI and assistive technology. In addition to virtual coaching programs that offer advice on symptom management and health improvement, this can also involve wearable technology that monitors vital signs and provides feedback on health behaviors.
(d) Digital care: In locations with limited access to health care, AI and assistive technology can be employed to give women virtual care. Women may obtain treatment while relaxing in their own homes thanks to telemedicine consultations, online support networks, and remote monitoring tools.

16.5 Women's health care and AI: Trust issues

Being able to trust AI and ML is a crucial problem since these technologies are being utilized more and more in women's health care. Although AI and ML offer a great deal of promise to enhance the lives of women, trusting these technologies also comes with several risks (Fig. 16.6). The primary difficulties are as follows:

1. Insufficient transparency: One of the biggest obstacles towards embracing AI and ML in women's health care is the opaque nature of these systems' inner workings. It can be challenging for women consumers to comprehend how judgments are being made since numerous algorithms for ML are complicated and opaque. The absence of transparency can undermine the already low confidence of women and make it challenging to confirm the system's correctness and fairness [36].

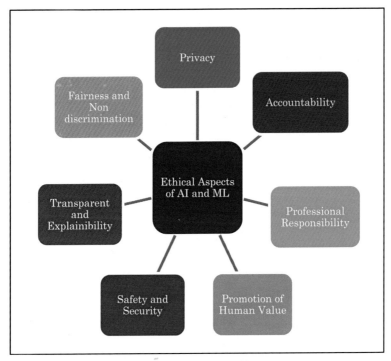

FIG. 16.6   Ethical aspects of AI and ML in women's health care.

2. Reliability: As AI and ML technologies are increasingly employed in women's health care where accuracy and dependability are crucial, reliability of these systems is a major problem. Reliable outcomes require the use of superior information that is typical of the women and their physiological and biological health. In absence of rational justification, women fail to understand the reasoning behind the system's decisions and do not rely on the use of AI and ML to a full extent in their health care [37,38].

3. Discrimination and bias: If AI and ML systems are built to reinforce preexisting societal biases or if they receive instruction on input that is not typical of the community, they may be biased. The designing and building of AI systems does not consider the specific needs of women and their health concerns, which are different than those in men. For example, if an algorithm is designed to predict future onset of heart disease based on past data, but that data contains biases, the algorithm can replicate those biases in its predictions. Women are at a less risk of heart ailments but at greater risk of other health anomalies, which AI and ML will not capture, thus the risk assessment will be flawed [39].

4. Limited accountability: It is challenging to hold AI and ML systems responsible for their decisions because, unlike human decision-makers, they lack a

moral compass. This may result in circumstances when individuals are injured by automated judgments and are unable to contest them. Because of their intricacy and the brisk rate of invention in these domains, regulation of AI and ML may be difficult. This may make it challenging to define precise rules for their proper usage, which may result in a lack of responsibility [40].

5. Cybersecurity risks: Systems using AI and ML may be subject to cyberattacks, which can make women less inclined to use technology and destroy their faith in it. Attackers may hack the system to acquire private information or obstruct vital processes. The data on women's health care is deemed very private and sensitive, especially data related to their reproductive health, pregnancy, abortion, and other cosmetic conditions. If women do not trust the system, they will not use the system for sensitive health conditions [41].

6. Ethical considerations: Privacy issues arise with the collecting and employing of personal data, especially when the data is critical or individually identifiable as in the case of women's health records. AI systems' ability to make decisions on their own poses issues with accountability and responsibility. In case of any misuse and complication, there is no one liable. Women need assurance and convincing for making health decisions for themselves and their families and this will be difficult to obtain from AI. The question will remain, who is responsible for the consequences of the choices that autonomous AI systems make? [42–44].

Overall, overcoming these difficulties through accountability, openness, fairness, and ethical concerns is necessary to increase confidence in AI and ML for women's health care. We can make sure that these innovations are applied in ways that serve women and their specific health requirements.

16.6 Ways to mitigate trust and bias

1. Examine the methodology and data to identify any areas where a serious risk of unfairness exists. As an example: Verify the training dataset's size and representativeness to prevent biases like random bias. Subpopulation analysis is the process of computing system metrics for different dataset subsets. This can help determine whether the performance of the model remains stable across subpopulations. Always keep an eye out for biases in the model. The outcomes may change as ML algorithms improve or as learning data changes [45].

2. Construct a debiasing strategy that incorporates several organizational, operational, and technological activities into your entire AI strategy: Technical approach refers to the application of techniques to locate possible prejudices sources and pinpoint aspects of the data that affect the model's correctness. One of the operational strategies is to use internal "red teams"

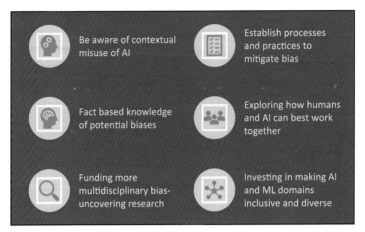

FIG. 16.7   Steps to minimize bias and increase women trust in the AI and ML healthcare system.

and independent auditors to enhance data collection processes. Other methodologies are used in Google AI's study on fairness. Organizational strategy includes creating a workplace where metrics and methods are openly shared [46].

3. As biases in instruction data are found, human-driven processes have to be improved. Through model creation and evaluation, biases that have been hidden for a while may come to light. When creating AI models, businesses may be aware of these prejudices and utilize this knowledge to understand the root reasons of prejudice. By instruction, method development, and cultural adjustments, businesses may improve the process to reduce bias [47].

4. Establish which cases call for human intervention and which are better served by computerized decision-making [17].

5. Create an organization that is varied. Due to the variety of the AI community, biases are easier to identify. Users who are members of that specific minority group are more inclined to raise concerns about bias at first. Keeping an inclusive AI team can therefore help to lessen inadvertent AI biases [48].

Fig. 16.7 shows the six steps McKinsey has devised to help practitioners of AI and ML to minimize bias [49]. From awareness creation to funding and investing in bias and trust related-research, AI and ML systems require a multi-pronged approach to tackle the ethical dimensions and build trust of women in automation and advanced machinations.

16.7   Conclusion

New care delivery methods, particularly those that may be offered by AI, are still not widely believed to be effective and secure. As AI gradually permeates the medical industry, it should provide competent, empathetic, and ethically responsible care for individuals. Because they are either unaware of them or do not have recourse to them, women may be less inclined to employ AI and ML technologies. Due to a lack of engagement in the creation and application of these technologies, gender biases may persist, and less efficient systems may result.

It is crucial to take into account aspects like cultural norms, language difficulties, and accessibility to technology in order to make sure that AI and ML techniques are employed to offer equal care for all women. Incorporating the special requirements and perspectives of women from all backgrounds into the development and implementation of these technologies is also crucial. This implies that AI should be built using impartial data and algorithms and should be geared to serve the unique demands of varied groups of women. When integrating AI in health care, it is also crucial to preserve women's privacy.

In particular, if consumers are given the ability to ask questions about and engage with how firms use their information and are given clear answers with a stronger, more overt cooperation between industry and academia, it will change how consumers see health care. The industry will be required to give openness and hence build trust by involving the public and respectable medical investigators at the beginning of the study process. This will enhance the two qualities—openness and trust—that are necessary for healthcare administrators to accept innovations into a wider healthcare system. Without trust, patients' perceptions of risk would continue to eclipse the potential of revolutionary ways to revolutionize health care.

References

[1] J. Abelson, F.A. Miller, M. Giacomini, What does it mean to trust a health system? A qualitative study of Canadian health care values, Health Policy (Amsterdam, Netherlands) 91 (1) (2009) 63–70, https://doi.org/10.1016/j.healthpol.2008.11.006.

[2] V. Alexander, C. Blinder, P.J. Zak, Why trust an algorithm? Performance, cognition, and neurophysiology, Comput. Hum. Behav. 89 (2018) 279–288, https://doi.org/10.1016/j.chb.2018.07.026.

[3] D. Castelvecchi, Can we open the black box of AI? Nature 538 (7623) (2016) 20–23, https://doi.org/10.1038/538020a.

[4] M. Ryan, In AI we trust: ethics, artificial intelligence, and reliability, Sci. Eng. Ethics 26 (2020) 2749–2767.

[5] E. Schniter, T.W. Shields, D. Sznycer, Trust in humans and robots: economically similar but emotionally different, J. Econ. Psychol. 78 (2020) 102253, https://doi.org/10.1016/j.joep.2020.102253.

[6] Z. Obermeyer, B. Powers, C. Vogeli, S. Mullainathan, Dissecting racial bias in an algorithm used to manage the health of populations, Science 366 (6464) (2019) 447–453.

[7] M. Renner, S. Lins, M. Söllner, S. Thiebes, A. Sunyaev, Understanding the necessary conditions of multi-source trust transfer in artificial intelligence, in: Proceedings of the 55th Hawaii International Conference on System Sciences (HICSS2022), 2022, pp. 1–10. http://hdl.handle.net/10125/80057.

[8] C.P. Langlotz, Will artificial intelligence replace radiologists? Radiol. Artif. Intell. 1 (3) (2019) e190058, https://doi.org/10.1148/ryai.2019190058.

[9] H. Nissenbaum, Accountability in a computerized society, Sci. Eng. Ethics 2 (1) (1996) 25–42, https://doi.org/10.1007/BF02639315.

[10] A. Adjekum, M. Ienca, E. Vayena, What is trust? Ethics and risk governance in precision medicine and predictive analytics, Omics: J. Integr. Biol. 21 (12) (2017) 704–710, https://doi.org/10.1089/omi.2017.0156.

[11] P. Ruotsalainen, B. Blobel, A model for calculated privacy and trust in pHealth ecosystems, Stud. Health Technol. Inform. 249 (2018) 29–37.

[12] M.A. Hall, E. Dugan, B. Zheng, A.K. Mishra, Trust in physicians and medical institutions: what is it, can it be measured, and does it matter? Milbank Quart. 79 (4) (2001) 613–639, https://doi.org/10.1111/1468-0009.00223.

[13] M. Wolfensberger, A. Wrigley, Trust in Medicine: Its Nature, Justification, Significance, and Decline, Cambridge University Press, 2019.

[14] M. Oshana, Trust and autonomous agency, Res. Philos. 91 (3) (2014) 431–447.

[15] R. Lukyanenko, W. Maass, V.C. Storey, Trust in artificial intelligence: from a foundational trust framework to emerging research opportunities, Electr. Mark. 32 (2022) 1993–2020, https://doi.org/10.1007/s12525-022-00605-4.

[16] T. Yamagishi, Trust: The Evolutionary Game of Mind and Society, Springer Tokyo, 2011.

[17] K. Siau, W. Wang, Building trust in artificial intelligence, machine learning, and robotics, Cutter Bus. Technol. J. 31 (2018) 47–53.

[18] O. Asan, A.E. Bayrak, A. Choudhury, Artificial intelligence and human Trust in Healthcare: focus on clinicians, J. Med. Internet Res. 22 (6) (2020) e15154, https://doi.org/10.2196/15154.

[19] K. Joyce, L. Smith-Doerr, S. Alegria, S. Bell, T. Cruz, S.G. Hoffman, S.U. Noble, B. Shestakofsky, Toward the sociology of artificial intelligence: a call for research on inequalities and structural change, Socius 7 (2021) 1–11, https://doi.org/10.1177/2378023121999581.

[20] J. Anderson, L. Rainie, Artificial Intelligence and the Future of Humans, Pew Research Center, 2018. https://www.pewresearch.org/internet/2018/12/10/artificial-intelligence-and-the-future-of-humans/.

[21] P. Nickel, Trust in medical artificial intelligence: a discretionary account, Ethics Inf. Technol. 24 (2022), https://doi.org/10.1007/s10676-022-09630-5.

[22] M. Nagendran, Y. Chen, C.A. Lovejoy, A.C. Gordon, M. Komorowski, H. Harvey, et al., Artificial intelligence versus clinicians: systematic review of design, reporting standards, and claims of deep learning studies, BMJ 368 (2020) m689, https://doi.org/10.1136/bmj.m689.

[23] M. Calnan, R. Rowe, Trust and health care, Sociol. Compass 1 (2007) 283–308.

[24] University of Tokyo, Measuring Trust in AI: Researchers Find Public Trust in AI Varies Greatly Depending on the Application, ScienceDaily, 2022. www.sciencedaily.com/releases/2022/01/220110103323.htm.

[25] Y. Ikkatai, T. Hartwig, N. Takanashi, H.M. Yokoyama, Octagon measurement: public attitudes toward AI ethics, Int. J. Hum.–Comput. Interact. 38 (17) (2022) 1589–1606, https://doi.org/10.1080/10447318.2021.2009669.

[26] B.K. Petersen, J. Chowhan, G.B. Cooke, R. Gosine, P.J. Warrian, Automation and the future of work: an intersectional study of the role of human capital, income, gender and visible minority status, Econ. Ind. Democr. (2022), https://doi.org/10.1177/0143831X221088301.

[27] N. Mehrabi, F. Morstatter, N. Saxena, K. Lerman, A. Galstyan, A survey on bias and fairness in machine learning, ACM Comput. Surveys (CSUR) 54 (6) (2021) 1–35, https://doi.org/10.1145/3457607.

[28] E.J. Topol, High-performance medicine: the convergence of human and artificial intelligence, Nat. Med. 25 (1) (2019) 44–56, https://doi.org/10.1038/s41591-018-0300-7.

[29] K.H. Yu, A.L. Beam, I.S. Kohane, Artificial intelligence in healthcare, Nat. Biomed. Eng. 2 (10) (2018) 719–731, https://doi.org/10.1038/s41551-018-0305-z.

[30] D. Lee, S.N. Yoon, Application of artificial intelligence-based technologies in the healthcare industry: opportunities and challenges, Int. J. Environ. Res. Public Health 18 (1) (2021) 271, https://doi.org/10.3390/ijerph18010271.

[31] M. Miyashita, M. Brady, The health care benefits of combining wearables and AI, Harv. Bus. Rev. (2019). Available Online: https://hbr.org/2019/05/the-health-care-benefits-of-combining-wearables-and-ai. (Accessed 18 June 2020).

[32] R. Kaur, R. Kumar, M. Gupta, Predicting risk of obesity and meal planning to reduce the obese in adulthood using artificial intelligence, Endocrine 78 (3) (2022) 458–469, https://doi.org/10.1007/s12020-022-03215-4.

[33] R. Kaur, R. Kumar, M. Gupta, Food image-based nutritional management system to overcome polycystic ovary syndrome using deep learning: a systematic review, Int. J. Image Graph. (2022) 2350043.

[34] R. Kaur, R. Kumar, M. Gupta, Deep neural network for food image classification and nutrient identification: a systematic review, Rev. Endocr. Metab. Disord. (2023), https://doi.org/10.1007/s11154-023-09795-4.

[35] R. Kaur, R. Kumar, M. Gupta, Food image-based diet recommendation framework to overcome PCOS problem in women using deep convolutional neural network, Comput. Electr. Eng. 103 (2022) 108298.

[36] J. Guo, B. Li, The application of medical artificial intelligence technology in rural areas of developing countries, Health Equity 2 (1) (2018) 174–181, https://doi.org/10.1089/heq.2018.0037.

[37] D.S. Kermany, M. Goldbaum, W. Cai, C.C.S. Valentim, H. Liang, S.L. Baxter, A. McKeown, G. Yang, X. Wu, F. Yan, J. Dong, M.K. Prasadha, J. Pei, M.Y.L. Ting, J. Zhu, C. Li, S. Hewett, J. Dong, I. Ziyar, A. Shi, et al., Identifying medical diagnoses and treatable diseases by image-based deep learning, Cell 172 (5) (2018), https://doi.org/10.1016/j.cell.2018.02.010. 1122–1131.e9.

[38] M. Sutrop, Should we trust artificial intelligence? Trames 23 (2019) 499–522.

[39] J. Tallant, You can trust the ladder, but you shouldn't, Theoria (2019), https://doi.org/10.1111/theo.12177.

[40] S.A. Voerman, P.J. Nickel, Sound trust and the ethics of telecare, J. Med. Philos. 42 (1) (2017) 33–49, https://doi.org/10.1093/jmp/jhw035.

[41] J. Wanner, L.V. Herm, K. Heinrich, The effect of transparency and trust on intelligent system acceptance: evidence from a user-based study, Electr. Mark. 32 (4) (2022) 2079–2102, https://doi.org/10.1007/s12525-022-00593-5.

[42] M. Taddeo, On the risks of trusting artificial intelligence: the case of cybersecurity, in: J. Cowls, J. Morley (Eds.), The 2020 Yearbook of the Digital Ethics Lab. Digital Ethics Lab Yearbook, Springer, 2021, pp. 97–108, https://doi.org/10.1007/978-3-030-80083-3_10.

[43] I. Saif, B. Ammanath, Trustworthy AI Is a Framework to Help Manage Unique Risk, MIT Technology Review, 2020. Retrieved February 28, 2023 from https://www.technologyreview.com/2020/03/25/950291/trustworthy-ai-is-a-framework-to-help-manage-unique-risk/.

[44] D.S. Char, N.H. Shah, D. Magnus, Implementing machine learning in health care—addressing ethical challenges, N. Engl. J. Med. 378 (11) (2018) 981–983, https://doi.org/10.1056/NEJMp1714229.

[45] C. Macrae, Governing the safety of artificial intelligence in healthcare, BMJ Qual. Safety 28 (6) (2019) 495–498, https://doi.org/10.1136/bmjqs-2019-009484.

[46] V. Polonski, People Don't Trust AI: Here's How We Can Change That, The Conversation, 2018. Retrieved March 27, 2023 from https://theconversation.com/people-dont-trust-ai-heres-how-we-can-change-that-87129.

[47] J. Shaw, F. Rudzicz, T. Jamieson, A. Goldfarb, Artificial intelligence and the implementation challenge, J. Med. Internet Res. 21 (7) (2019) e13659, https://doi.org/10.2196/13659.

[48] ELSI Advisory Group, Ethical Framework for Responsible Data Processing in the Swiss Personalized Health Network, Bern, 2017.

[49] F. Rossi, Building trust in artificial intelligence, J. Int. Aff. 72 (1) (2018) 127–134. https://jia.sipa.columbia.edu/building-trust-artificial-intelligence.

Chapter 17

Role of artificial intelligence and machine learning in women's health

Sapna Rawat[a], Poonam Joshi[b], Gulafshan Praveen[c], and Jyoti Saxena[a]
[a]*JBIT Group of Institution, Dehradun, Uttarakhand, India,* [b]*Uttaranchal Institute of Pharmaceutical Sciences, Uttaranchal University, Dehradun, Uttarakhand, India,* [c]*Veer Madho Singh Bhandari Uttarakhand Technical University, Dehradun, Uttarakhand, India*

17.1 Introduction

Artificial intelligence (AI) has significantly impacted science in recent years. Because AI has numerous effective uses in the medical industry, specialists in the field must understand AI's fundamental concepts. AI systems are defined in the European Commission's communications. [1,2].

The study of AI combines computer science, mathematics, and engineering [3]. Research has shown that there is a gender bias in science, technology, engineering, and mathematics (STEM) fields, and that this masculine bias has an impact on how intelligent systems are built. In the realm of AI, women conduct fascinating research but are frequently marginalized for a variety of reasons. It is crucial to take the initiative to promote the vital work done by female scientists in the many STEM fields, particularly in AI. AI is becoming more and more significant in the development of emerging digital society [4].

Data analysis, information, and the decision-making process are all significantly influenced by AI, and AI is already improving some parts of health care. According to Deloitte's State of AI survey, mid-sized organizations invest less in AI projects and innovations than major corporations, who collectively spend more than US $50 million. The three main goals that healthcare businesses are pursuing using AI are cost reduction, process improvement, and enhancement of current products and services. The present pandemic has recognized the significance of AI and provided a platform for digital technologies like AI to address issues [5].

Artificial Intelligence *and* Machine Learning *for* Women's Health Issues
https://doi.org/10.1016/B978-0-443-21889-7.00006-3

The use of AI and machine learning (ML) to explain data and automate time-consuming and expensive tasks is gaining popularity in women's health care. AI and ML have the potential to speed up decision making for operational and clinical staff, drastically reduce time spent on administrative work, and free up human attention for more difficult, meaningful, and important managerial and clinical work. Anywhere along the healthcare value chain, Deloitte's AI-driven, cloud-based intellectual therapy can be used toward progress in working efficiency, lowering costs, and providing enhanced consumer health outcomes [6].

17.1.1 Artificial intelligence and women's health

AI is being utilized in medicine to forecast patient outcomes and offer real-time data about health concerns. It falls into two major types.

- Methods for collecting data from unstructured sources using natural language
- ML techniques for analyzing structured data [7]

AI is used in medicine in at least four different ways: to assess the probability that a disease will develop, to predict treatment success, to control or minimize consequences, and to identify the pathophysiology of a disease and the best course of therapy [8]. The focus of women's health in AI research has been on the cardiovascular system, the breasts, the cervix, the skeletal system, and the endometrium.

AI has been used to predict endometrial cancer in postmenopausal women, calculate the role of different categories of human papillomavirus (HPV) in the recurrence of cervical intraepithelial neoplasia, and enhance breast imaging diagnostics. Popular ML methods used to create predictive models for osteoporosis include random forest (RF), support vector machine (SVM), artificial neural networks (ANNs), and logistic regression (LR). One study compared the ability of different ML systems to predict fragility fractures using MRI data. In the treatment of people suffering from hormonal and metabolic disorders, AI has been utilized to determine bone age. Promising outcomes have been obtained when AI technologies have been used to interpret statistics from stroke imaging [10]. Prognostic categorization models have been used to create decision-making systems based on ML to forecast outcomes in particular cases. AI has additionally been employed in preventative medicine to identify people who are more likely to develop heart disease, bowel cancer, colon cancer, and other problems. Applying ML techniques to clinical concerns has been significantly hampered by the dearth of adequately labeled medical data [10,11].

17.2 Aim

The following concerns are discussed in this chapter:

- The definitions of deep learning (DL), ML, and AI
- AI in health diagnostics and opportunities and issues for clinical practice

- The significance of AI in women's health care
- The effects of AI on the working lives of women

What are DL, ML, and AI?
Merriam-Webster defines AI as:

1. A division of computer science focused on modeling intelligent conduct in computers
2. Ability of a device to simulate smart human action

Wikipedia defines AI as follows:
"AI, which is different from human intelligence, usually referred to as knowing about machine intelligence in computer science, is knowledge produced by computers. According to prominent AI textbooks the study of 'intelligent agents', or any kind of technology which observes surroundings that behaves in a manner that optimizes the chance of accomplishing its goals computers that mimic 'perceptive' methods that individuals generally subordinate through conscious imagination, for example 'knowledge' and 'problem-solving', are sometimes referred to as 'artificial intelligence' in popular language [12]."

17.2.1   Presenting AI in terms of levels makes it easier to comprehend

The initial stage of AI is represented by easy-to-control software that is operated by the worker directly. Another stage of AI is classical, through many input–output mappings and different patterns. These systems are used when looking, estimating, and databases of knowledge. The third stage of AI is based on ML. The fourth stage of AI is based on DL. The area of ML, it is claimed, gives computers the ability to learn without explicit programming (Fig. 17.1).

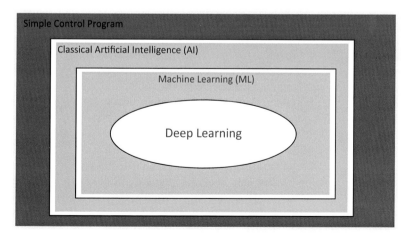

FIG. 17.1   The dimensions of artificial intelligence.

Machine learning is defined as "the scientific investigation of algorithms as well as statistical models used by computer systems to perform a specific task without human intervention," which is given by Samuel's in 1959 [13]. ANNs used in DL have numerous hidden units, hence the term "deep" neural networks. DL is a method for ML.

17.3 Role of artificial intelligence in science

As AI can improve scientific and research productivity, it is becoming more and more significant in the field. Also, it can promote the reproducibility of scientific research and provide new opportunities for discovery. Applications for AI include the evaluation of huge datasets, the creation of theories, the comprehension and analysis of scientific literature, and the facilitation of data collection, experimental design, and experimentation. Improvements in computer technology and software, the accessibility of data, and the availability of open-source AI code are some recent drivers of AI in science. AI is being used more and more in scientific applications, including automated ML processes, hypothesis development, reviews, interpretation, and analysis of scientific literature [14–16].

17.3.1 Role of artificial intelligence in health care

Applications of AI in medical service can aid in early disease detection, the provision of preventative services, the optimization of clinical decision-making, and the development of novel therapies and drugs. They can also support self-regulatory devices, programs, and tools, along with specific health care and precision medicine. AI in medicine has the potential to save costs and improve service quality, but it does have ethical impacts on data protection. This section focuses on the effects of AI on health care. AI-enhanced solutions can spot harmful behaviors and promote positive ones, giving real-time feedback throughout the care continuum. This can control and enhance the health system's effectiveness and efficiency. AI systems can reduce costs by identifying redundant services, underutilized opportunities, and inefficient therapies. They can also make sure that patients receive the services that are best suited to their needs, accurately predict future healthcare needs, and optimize resource distribution throughout the system. Data-driven innovation aims to create a "learning health system" that continuously mixes data from researchers, providers, and patients [17].

The detection and dissemination of public health information, the facilitation of health research, and the diagnosis of medical disorders are all made possible by AI technologies. Big data analytics provides fresh ways to monitor health and disease progression for more accurate diagnosis and management [18]. Moreover, AI is being utilized to train models to categorize photos of the eye, with the goal of integrating cataract detectors into cell phones and transporting them to isolated regions. To fully utilize AI in health care, the necessary infrastructure and risk management measures need be in place. Although more nations are implementing electronic health record (her) systems and embracing mobile health, only a small number have attained high-level integration. To fully utilize EHRs, standards and interoperability are essential obstacles to overcome, as is reducing privacy hazards. Examples include the utilization of data gathered from biological samples and implantable medical devices for ML [19,20].

17.4 Importance of artificial intelligence (AI) in health care and medicine for women

Researchers have investigated the use of ANNs, classification, and regression trees to predict endometrial cancer in postmenopausal women. Similar to this, AI is used to calculated how different HPV types affect the likelihood of a recurrence of cervical dysplasia. In breast ultrasound, where data from mammography, solography, and magnetic resonance imaging (MRI) were obtained for studies, AI has been extensively applied [21,22].

Congenital heart disease risk among pregnant women has been predicted using ANNs. It was discovered that the models could identify patients early in pregnancy who were most susceptible to severe cardiac disease with healthy born predictions, potential for implantation of fertilized ovum, the effect of endometriosis on the results of technologies for assisted reproduction, and other factors. It has been argued that AI could significantly improve the personalization of treatments for infertility [11,23].

From radiography to chronic diseases like cancer, AI is being pushed to develop precise and effective products that will treat patients and, ideally, find a cure for these diseases [24].

17.4.1 Opportunities and issues for medical care with AI in health diagnostics

Significant evidence of the most possible uses of AI techniques using DL research for application in the diagnosis of illnesses has been exposed. While fundamental investigation into these methods should continue to advance, we advocate parallel, focused work on emerging comprehensive evaluation and techniques for clinically validating AI algorithms. This research should aim to improve assurance among medical professionals, give insight into the specific areas that need improvement, and identify and resolve any implementation issues as soon as they arise [25–28].

17.5 Using MI to predict complications in pregnancy

ML is gaining importance in health care, with usefulness for diagnosis and prediction, as well as for setting priorities. In 2017, almost 295,000 women lost their lives during, immediately during, or soon after childbirth. It is understood that perinatal problems can best be predicted and avoided during the first trimester of pregnancy.

Two of the perinatal conditions that may be predicted during pregnancy with the help of ML are prematurity (seven studies) and preeclampsia (six studies), and these have been divided into 15 main prediction outputs:

1. prematurity
2. preeclampsia
3. poor gestational size for delivery
4. gestational diabetes mellitus (GDM)
5. perinatal mortality
6. fetal academia
7. fetal hypoxia
8. placental accreta
9. pulmonary disease
10. Cesarean section
11. placental development
12. congenital defect
13. acute maternal disease
14. spontaneous abortion
15. trial of labor after caesarean (TOLAC)

17.5.1 The women health initiative: Predicting incident of heart failure in women with machine learning

Digital health records, the number of which are growing due to the frequency of medical tests, are unstructured. Wearable devices are being utilized to monitor health parameters to diagnose cardiac issues. However, data from wearable sensors to detect heart illness can be contaminated by aberrations such as unreported values and noise, which reduce efficiency and misreport outcomes [29]. ML is extremely vast and sophisticated, and both its scope and applications are constantly growing. ML uses a range of classifiers from supervised, unsupervised, and ensemble learning to forecast and assess the correctness of the provided dataset. It is crucial for patients, their healthcare practitioners, healthcare systems, and third-party payers to be able to predict mortality in heart failure (HF). However, it has shown to be challenging to reliably quantify outcomes in patients with HF. Several methods, such as biomarkers, risk scores, and their combinations, have been developed for this goal, however, the majority of them are ineffective, especially when used on HF populations different from the ones for whom the scores were created [30].

There are different risk factors and heart failure (HF) risk are present. Conventional methods for HF risk prediction omit crucial elements like cardiac biomarkers [31]. Australia's biggest cause of mortality for both men and women is cardiovascular disease, with clear disparities in prognosis and treatment. In addition, the pathophysiology of disease differs between men and women. With a lower ejection fraction, men are more prone to develop macrovascular disease or HF. Women are more likely to experience endothelial dysfunction and microvascular coronary problems coupled with HF with a maintained ejection fraction. There is evidence to support the notion that women are frequently misdiagnosed and mistreated for both conventional and new cardiovascular risk factors. Additionally, several "conventional" risk factors, such as excessive smoking, may indicate that women are more likely than men to develop cardiovascular disease [32,33].

17.5.2   Identifying novel female-specific predictors

HF with preserved ejection fraction (HFpEF) is a syndrome caused by various factors, including long-term inflammation, endothelial dysfunction, and numerous tissue problems. The primary last stage of atherosclerosis is HF with reduced ejection fraction (HFrEF). Even the most recent pharmacological treatments for HFpEF were not developed for HFpEF particularly; instead, they were first meant to treat HFrEF or diabetes and were later repositioned for HFpEF. Endothelial dysfunction and coronary microvascular disease may play important roles in the development of cardiovascular pathology in HFpEF, according to an increasing body of evidence, with dysregulated inflammation being a significant contributor to this pathophysiology.

There are no existing remedies that explicitly mark irritation or pathways involved to prevent HFpEF. The multi-ethnic study of atherosclerosis (MESA) was created to evaluate patients who had subclinical cardiovascular disease at the beginning of the trial and to determine whether they later developed cardiovascular disease and experienced events associated with cardiovascular disease. In the MESA trial, HF was a judged endpoint. Participants were classified as having HFpEF if their adjudicated endpoint was HF with at least a 45% ejection fraction.

Using MESA data, interleukin-2 (IL-2) was ranked according to its association with newly diagnosed cardiac disease. The study did not make a distinction between HFpEF and HFrEF, however. Additionally, a link between IL-2 and HF was discovered. Currently, hypothesis-driven research does not alter data; therefore IL-2 is associated with HFpEF but not HFrEF. Hence assisting in the differentiation of the diverse pathophysiology, which are primarily separated into HFpEF and HFrEF [9,33,34].

17.6 The effect of AI on working lives of women

AI and automation can lead to increased efficiency, productivity, and job creation in certain industries, which can benefit women who work in those fields [35]. For example, AI can help reduce the gender pay gap by automating routine tasks and providing women with more opportunities for higher-skilled and higher-paying jobs. On the other hand, there are concerns that AI and automation could also have negative impacts on women's working lives. As an illustration, there is evidence that women may be disproportionately affected by job displacement and wage stagnation because of automation. In addition, there are concerns that AI and ML algorithms may perpetuate gender biases in hiring and promotion decisions, further exacerbating existing inequalities in the workplace. Furthermore, women may face unique challenges in the growth and adoption of AI tools due to the lack of diversity in the tech industry. The underrepresentation of women and other marginalized groups in tech can lead to the development of AI systems that do not adequately consider the needs and experiences of diverse populations, potentially perpetuating biases and discrimination [36]. In summary, the impact of AI on the working lives of women is complex and depends on various factors, including the nature of the job and the industry, the level of diversity in the tech industry, and the extent to which biases are addressed in the growth of AI skills.

Women's employment prospects, as well as their standing, status, and treatment at work, will be impacted by the adoption of AI technologies [37]. Women in the workforce earn less than males do globally, spend more time giving children and the elderly unpaid care, have fewer senior positions, participate in fewer STEM industries, and generally have more unstable professions. The workplace is undergoing fast change. AI usage is increasing in both residential and professional contexts. Even though the percentage of women in the labor force increased globally in the 20th century, there is still much to be done to achieve gender equality in the workplace and elsewhere. Governments, organizations, and businesses must eliminate gender gaps rather than allowing them to persist or worsen through utilizing AI. AI can be applied by healthcare practitioners to progress patient care. The fact that AI algorithms are based on previous diagnostic data, which frequently includes information obtained from men, must also be considered [38].

17.6.1 Women in AI: Challenges

Women in AI face several challenges, which we discuss in the sections that follow.

6.1.1. Underrepresentation: Women are significantly underrepresented in the AI field, both in terms of employment and leadership positions. This lack of representation can lead to a lack of diversity in ideas, perspectives, and solutions.

6.1.2. **Bias:** AI algorithms and systems can be biased, and this bias can have a disproportionate impact on women and other underrepresented groups.

6.1.3. **Career progression:** Women in AI may face barriers to career progression, such as a lack of mentorship, limited networking opportunities, and unconscious bias in hiring and promotion decisions.

Imposter syndrome: Women in AI may experience imposter syndrome, a feeling of not belonging or being unworthy of their position.

6.1.4. **Work–life balance:** Like many other fields, AI can be demanding and require long hours and intense focus. Women may find it difficult to balance their work with other responsibilities, such as caring for children or elderly relatives.

Think about the real-world illustration of helmets, headrests, and airbags in cars, which were mostly designed using data from vehicle crash dummy experiments using men's bodies and their seating positions. The "standard" measures do not take into account women's breasts or pregnant bodies [38,39]. Women are consequently 17% more likely to die in a similar event and 47% more likely to have useful insurance than males.

The disparities between men's and women's occupations begins early, during the choosing of their academic specializations, according to the Organization for Economic Co-operation and Development (OECD; 2017). For instance, compared to 5% of boys, on average, just 0.5% of 15-year-old girls in OECD nations want to work in the ICT industry. In the domains of STEM, boys are twice as likely as girls to plan to pursue careers as scientists, engineers, or architects.

To address these challenges, it is important to increase diversity and representation in the AI field, provide opportunities for mentorship and networking, promote transparency and accountability in AI algorithms and systems, and support work-life balance and wellbeing for women in AI.

17.6.2 The impact of AI-driven automation on the work of women

The impact of AI-driven automation on the work of women is a topic of growing concern, as there are both potential benefits and drawbacks to consider.

However, there are also concerns that AI-driven automation could lead to job displacement and wage stagnation, particularly for women who are overrepresented in regular tasks and low-skilled work that are more likely to be automated. Furthermore, the implementation of AI-driven automation could require women to acquire new skills and knowledge, which may not be accessible to all women due to factors such as cost or educational barriers. This could lead to a widening skills gap between women and men, and potentially exacerbate existing gender inequalities in the workplace.

If automation has the same impact as previous technological shocks, both men and women may experience employment loss and prospective job growth of about comparable magnitude. By 2030, the average number of working women in the 10 countries under study (107 million) could lose their employment to automation, compared to the average number of males (163 million). If the demand for labor increases as expected, women may have access to 20% more jobs than males, according to estimates based on their respective proportions in industries and occupations. There will also be whole new occupations established, but in the United States, about 60% of these occupations have been in fields where men predominate.

In summary, while AI-driven automation can bring benefits to the work of women, there are also potential drawbacks that should be properly considered and addressed. Setting priorities is crucial, as is diversity inclusion in the development and implementation of AI-driven automation, and support of women in acquiring the skills and knowledge needed to adapt to a rapidly changing work environment [40].

17.7 Conclusion

AI has had a significant effect on the sciences, particularly in health care. It is being used in medicine to forecast patient outcomes and offer real-time data about health concerns. The provision of health care to women is seeing a rise in the use of algorithms, ML, and AI to identify and prevent maternal health problems such as premature birth, proteinuria, undesirable delivery, GDM, infant mortality, fetal academia, placental accrete, pulmonary disease, caesarean section, significant maternal morbidity, spontaneous abortion, and congenital abnormality. ML is used to forecast mortality due to HF. This study focused on two classifiers, LR and SVM, which had much lower HF discrimination against women's Health Initiative WHI. This demonstrates that an ML-based strategy may be advantageous for populations whose predictors may not correspond to current risk models. AI-driven automation has the ability to minimize the load of routine and routine work, making time available for women to concentrate on better planning and strategic activity. To address the challenges of AI in women's health care, it is important to increase diversity and representation, provide opportunities for mentorship and networking, promote transparency and accountability, and support work-life balance and wellbeing. By 2030, women may have access to 20% more jobs than males.

References

[1] A.C. Prietoa, Artificial Intelligence and Automation of Systematic Reviews in Women's Health, Copyright © 2020 Wolters Kluwer health, Inc. All rights reserved.page 337, 2020, https://doi.org/10.1097/GCO. www.co-obgyn.com.

[2] Commission. E, A Definition of Artificial Intelligence: Main Capabilities and Scientific Disciplines, 2019.

[3] V. Aida, G. Karina, Women in Artificial Intelligence, Appl. Sci. 12 (2022) 9639. https://doi.org/10.3390/app12199639. 2022.

[4] European Commission, Women in the Digital Age, Available online https://op.europa.eu/en/publication-detail/-/publication/84bd6dea-2351-11e8-ac73-01aa75ed71a1. Accessed on 1 September 2022.

[5] S. Samoili, M. López Cobo, B. Delipetrev, F. Martínez-Plumed, E. Gómez, G. De Prato, AI Watch. Defining Artificial Intelligence 2.0. Towards an Operational Definition and Taxonomy for the AI Landscape; EUR 30873 EN; Publications Office of the European Union: Luxembourg, ISBN 978-92-76-42648-6. [CrossRef], 2021.

[6] https://www2.deloitte.com/us/en/insights/industry/health-care/artificial-intelligence-in-health-care.html.

[7] F. Jiang, Y. Jiang, H. Zhi, et al., Artificial intelligence in healthcare: past, present and future, Stroke Vasc. Neurol. 2 (2017) 230 (2017).

[8] A. Becker, Artificial intelligence in medicine: what is it doing for us today? Health Policy Technol. 8 (2019) 198–205.

[9] 8.N.W. Carris, et al., Novel biomarkers of inflammation in heart failure with preserved ejection fraction: analysis from a large prospective cohort study, BMC Cardiovasc. Disord. 22 (1) (2022). Available at: https://doi.org/10.1186/s12872-022-02656-z.

[10] P. Vogiatzi, A. Pouliakis, C. Siristatidis, An artificial neural network for the prediction of assisted reproduction outcome, J. Assist. Reprod. Genet. 2019 (36) (2019) 1441–1448.

[11] C. Cui, S.S. Chou, L. Brattain, C.D. Lehman, A.E. Samir, Data engineering for machine learning in women's imaging and beyond, Am. J. Roentgenol. 213 (2019) 216–226.

[12] Merriam-Webster, Incorporated, Merriam-Webster Dictionary [Internet], Merriam-Webster, Springfield, MA, Available from: https://www.merriam-webster.com/ [cited 2020 Mar 2].

[13] https://en.wikipedia.org/wiki/Artificial_intelligence. (cited 2020 Mar 2).

[14] "Global Defence Spending to Hit Post-Cold War High", IHS Markit, , https://ihsmarkit.com/research-analysis/global-defence-spending-to-hit-post-coldwar-high-in-2018.html, (18 December 2018).

[15] OECD, Skills and Jobs in the Internet Economy, OECD Digital Economy Papers, No. 242, OECD Publishing, Paris, 2014. https://doi.org/10.1787/5jxvbrjm9bns-en.

[16] D. Wyllie, How 'Big Data' is Helping Law Enforcement, 2013, PoliceOne.Com, 20 August https://www.policeone.com/police-products/software/Data-Information-SharingSoftware/articles/6396543-How-Big-Data-is-helping-law-enforcement/.

[17] Canadian Institute for Health Information, Better Information for Improved Health: A Vision for Health System Use of Data in Canada, 2013, in collaboration with Canada Health Infoway http://www.cihi.ca/cihi-ext-portal/pdf/internet/hsu_vision_report_en.

[18] J. Kleinberg, et al., Human Decisions and Machine Predictions, 2017. NBER Working Paper, No. 23180.

[19] S. Brodmerkel, Dynamic pricing: retailers using artificial intelligence to predict top price you'll pay, ABC News (2017). http://www.abc.net.au/news/2017-06-27/dynamicpricing-retailers-using-artificial-intelligence/8638340.

[20] A. De Jesus, Augmented Reality Shopping and Artificial Intelligence – Near-Term Applications, Emerj, 2018. https://www.techemergence.com/augmented-realityshopping-and-artificial-intelligence/.

[21] S. Hamid, The Oppoutunities and Risks of AI in Medicine Healthcare, 2016.

[22] T. Panch, P. Sczolovits, R. Aton, Artificial Intelligence, Machine learning and health systems, J. Glob. Health 8 (2) (2018) 020303. (2018).

[23] P. Vogiatzi, A. Pouliakis, Siristadisc., An artificial neural network for the prediction of assisted reproduction outcome, J. Assist. Reprod. Genet. 36 (2019) 1441–1448.

[24] https://intellipaat.com/blog/artificial-intelligence-in-healthcare/.

[25] JASON, Perspectives on Research in Artificial Intelligence and Artificial General Intelligence Relevant to DoD, JSR-16-Task-003, 2017.

[26] A.L. Beam, I.S. Kohane, Translating artificial intelligence into clinical care, JAMA 316 (2016) 2368.

[27] C.S. Greene, et al., Opportunities and Obstacles for Deep Learning in Biology and Medicine, 2017, bioRxiv preprint first posted online https://doi.org/10.1101/142760.

[28] The Parable of Google Flu, D. Lazer, R. Kennedy, G. King, A. Vespignani, Traps in big data analysis, Science 343 (2014) 1203.

[29] F. Ali, et al., A smart healthcare monitoring system for heart disease prediction based on Ensemble Deep Learning and feature fusion, Inf. Fusion 63 (2020) 208–222. Available at https://doi.org/10.1016/j.inffus.2020.06.008.

[30] "Addendum to the article: 'improving risk prediction in heart failure using machine learning' [eur J heart fail 2020;22:139–147]" Eur. J. Heart Fail., 22(12), pp. 2399–2399. Available at: https://doi.org/10.1002/ejhf.2072 (2020).

[31] M.W. Segar, et al., Development and validation of machine learning–based race-specific models to predict 10-year risk of heart failure: a multicohort analysis, Circulation 143 (24) (2021) 2370–2383. Available at: https://doi.org/10.1161/circulationaha.120.053134.

[32] L. Geraghty, et al., Cardiovascular disease in women: from pathophysiology to novel and emerging risk factors, Heart Lung Circ. 30 (1) (2021) 9–17. Available at: https://doi.org/10.1016/j.hlc.2020.05.108.

[33] G.H. Tison, et al., Predicting incident heart failure in women with machine learning: the women's health initiative cohort, Can. J. Cardiol. 37 (11) (2021) 1708–1714. Available at https://doi.org/10.1016/j.cjca.2021.08.006.

[34] M.E. Hossain, S. Uddin, A. Khan, Network analytics and machine learning for predictive risk modelling of cardiovascular disease in patients with type 2 diabetes, Expert Syst. Appl. 164 (2021) 113918. Available at https://doi.org/10.1016/j.eswa.2020.113918.

[35] D. Acemoglu, P. Restrepo, Artificial intelligence, automation, and work, in: The Economics of Artificial Intelligence: An Agenda, University of Chicago Press, 2018, pp. 197–236.

[36] C. Collett, L.G. Gomes, G. Neff, The Effects of AI on the Working Lives of Women, UNESCO Publishing, 2022.

[37] A. Madgavkar, J. Manyika, M. Krishnan, K. Ellingrud, L. Yee, The Future of Women at Work, 2019.

[38] https://www.academia.edu/download/62326629/AI_Bias_Could_Put_Womens_Lives_At_Risk_-_A_Challenge_For_Regulators20200310-84471-17clytw.pdf.

[39] https://www.tandfonline.com/doi/abs/10.1080/08874417.2022.2089773.

[40] P.V. Moore, OSH and the future of work: benefits and risks of artificial intelligence tools in workplaces, in: Digital Human Modeling and Applications in Health, Safety, Ergonomics and Risk Management. Human Body and Motion: 10th International Conference, DHM 2019, Held as Part of the 21st HCI International Conference, HCII 2019, Orlando, FL, USA, Proceedings, Part I 21, Springer International Publishing, 2019, pp. 292–315. July 26–31.

Index

Note: Page numbers followed by *f* indicate figures, *t* indicate tables, and *b* indicate boxes.

Printed in the United States
by Baker & Taylor Publisher Services